AF411358

# Metal Oxides

# Metal Oxides

Editor

**Maria Luisa Grilli**

MDPI • Basel • Beijing • Wuhan • Barcelona • Belgrade • Manchester • Tokyo • Cluj • Tianjin

*Editor*
Maria Luisa Grilli
Italian National Agency for New Technologies,
Energy and Sustainable Economic Development,
Energy Technologies and Renewable Sources Department,
Casaccia Research Center
Italy

*Editorial Office*
MDPI
St. Alban-Anlage 66
4052 Basel, Switzerland

This is a reprint of articles from the Special Issue published online in the open access journal *Metals* (ISSN 2075-4701) (available at: https://www.mdpi.com/journal/metals/special_issues/metal_oxides).

For citation purposes, cite each article independently as indicated on the article page online and as indicated below:

LastName, A.A.; LastName, B.B.; LastName, C.C. Article Title. *Journal Name* **Year**, *Volume Number*, Page Range.

ISBN 978-3-03943-757-3 (Hbk)
ISBN 978-3-03943-758-0 (PDF)

# Contents


**About the Editor** . . . . . . . . . . . . . . . . . . . . . . . . . . . . . . . . . . . . . . . . . . . . . . . . **vii**

**Preface to "Metal Oxides"** . . . . . . . . . . . . . . . . . . . . . . . . . . . . . . . . . . . . . . . . . . **ix**

**Maria Luisa Grilli**
Metal Oxides
Reprinted from: *Metals* **2020**, *10*, 820, doi:10.3390/met10060820 . . . . . . . . . . . . . . . . . . . . . **1**

**Francesca D'Anna, Maria Luisa Grilli, Rita Petrucci and Marta Feroci**
$WO_3$ and Ionic Liquids: A Synergic Pair for Pollutant Gas Sensing and Desulfurization
Reprinted from: *Metals* **2020**, *10*, 475, doi:10.3390/met10040475 . . . . . . . . . . . . . . . . . . . . . **5**

**Adrian Mihail Motoc, Sorina Valsan, Anca Elena Slobozeanu, Mircea Corban, Daniele
Valerini, Mythili Prakasam, Mihail Botan, Valentin Dragut, Bogdan St. Vasile, Adrian Vasile
Surdu, Roxana Trusca, Maria Luisa Grilli and Robert Radu Piticescu**
Design, Fabrication, and Characterization of New Materials Based on Zirconia Doped with
Mixed Rare Earth Oxides: Review and First Experimental Results
Reprinted from: *Metals* **2020**, *10*, 746, doi:10.3390/met10060746 . . . . . . . . . . . . . . . . . . . . . **27**

**Mehmet Yilmaz, Maria Luisa Grilli and Guven Turgut**
A Bibliometric Analysis of the Publications on In Doped ZnO to be a Guide for Future Studies
Reprinted from: *Metals* **2020**, *10*, 598, doi:10.3390/met10050598 . . . . . . . . . . . . . . . . . . . . . **51**

**Hae-Jun Seok and Han-Ki Kim**
Study of Sputtered ITO Films on Flexible Invar Metal Foils for Curved Perovskite Solar Cells
Reprinted from: *Metals* **2019**, *9*, 120, doi:10.3390/met9020120 . . . . . . . . . . . . . . . . . . . . . **71**

**Sebastian Balos, Miroslav Dramicanin, Petar Janjatovic, Ivan Zabunov, Branka Pilic, Saurav
Goel and Magdalena Szutkowska**
Suppressing the Use of Critical Raw Materials in Joining of AISI 304 Stainless Steel Using
Activated Tungsten Inert Gas Welding
Reprinted from: *Metals* **2019**, *9*, 1187, doi:10.3390/met9111187 . . . . . . . . . . . . . . . . . . . . . **81**

**Sebastian Balos, Miroslav Dramicanin, Petar Janjatovic, Ivan Zabunov, Damjan Klobcar,
Matija Busic and Maria Luisa Grilli**
Metal Oxide Nanoparticle-Based Coating as a Catalyzer for A-TIG Welding: Critical Raw
Material Perspective
Reprinted from: *Metals* **2019**, *9*, 567, doi:10.3390/met9050567 . . . . . . . . . . . . . . . . . . . . . **95**

**Yuzhu Pan, Xuefeng She, Jingsong Wang and Yingli Liu**
Study of the Deposition Formation Mechanism in the Heat Exchanger System of RHF
Reprinted from: *Metals* **2019**, *9*, 443, doi:10.3390/met9040443 . . . . . . . . . . . . . . . . . . . . . **107**

**Magdalena Szutkowska, Sławomir Cygan, Marcin Podsiadło, Jolanta Laszkiewicz-Łukasik,
Jolanta Cyboroń and Andrzej Kalinka**
Properties of TiC and TiN Reinforced Alumina–Zirconia Composites Sintered with Spark
Plasma Technique
Reprinted from: *Metals* **2019**, *9*, 1220, doi:10.3390/met9111220 . . . . . . . . . . . . . . . . . . . . . **119**

**Sebastian Balos, Petar Janjatovic, Miroslav Dramicanin, Danka Labus Zlatanovic, Branka Pilic, Pavel Hanus and Lucyna Jaworska**
Microstructure, Microhardness, and Wear Properties of Cobalt Alloy Electrodes Coated with $TiO_2$ Nanoparticles
Reprinted from: *Metals* **2019**, *9*, 1186, doi:10.3390/met9111186 . . . . . . . . . . . . . . . . . . . . . 133

**Igor Luisetto, Maria Rita Mancini, Livia Della Seta, Rosa Chierchia, Giuseppina Vanga, Maria Luisa Grilli and Stefano Stendardo**
$CaO$–$CaZrO_3$ Mixed Oxides Prepared by Auto–Combustion for High Temperature $CO_2$ Capture: The Effect of CaO Content on Cycle Stability
Reprinted from: *Metals* **2020**, *10*, 750, doi:10.3390/met10060750 . . . . . . . . . . . . . . . . . . . . 143

# About the Editor

**Maria Luisa Grilli**, Ph.D., is a researcher at ENEA, the Italian National Agency for New Technologies, Energy and Sustainable Economic Development (Rome, Italy). She graduated in Physics at the University of Rome La Sapienza (1992) and received her Ph.D. in Materials Engineering at University of Rome Tor Vergata in 2001. She worked for an 11 month fellowship at the Laser Zentrum Hannover (Germany) in the frame of the Human Capital Mobility Network, Project "High Quality Thin Films for Laser Applications", and was awarded several national fellowships and post doc grants. Dr. Grilli worked from 1997 to 2005 at the Department of Chemical Science and Technology of University Tor Vergata, focusing on electrochemical gas sensors. She was an assistant professor of Material Technology and Applied Chemistry from 1999 to 2005. Her main research topics are design, fabrication (by PVD techniques) and the characterization of optical interference coatings for laser and space applications, R&D of coatings for thin film solar cells and optoelectronic devices, as well as the R&D of corrosion-resistant coatings (oxides, nitrides, reduced graphene oxide (GO prepared by the modified Hummer's method)). She has been involved in several EU, bilateral and national projects, and a co-organizer and speaker at international conferences. She was the coordinator of the Project of Particular Relevance Italy–China, "On-demand refractive index for remote sensing from space", WGs coordinator of the COST Action CA15102, "Solutions for Critical Raw Materials Under Extreme Conditions (CRM EXTREME)". She is an ENEA representative of the ERAMIN2 Project MONAMIX, "New concepts for efficient extraction of mixed rare earths oxides from monazite concentrates and their potential use as dopant in high temperature coatings and sintered materials" and of the Marie Skłodowska-Curie Actions ChemPGM "Chemistry of Platinum Group Metals". She is the chair of the COST INNOVATORS' Grant IG15102, ITHACA: "Innovative and sustainable technologies for reducing critical raw materials dependence for cleaner transportation applications". She is a peer reviewer for national European research proposals (Czech Republic), an external reviewer of COST Actions, and an active referee member of about 15 International Journals. She was co-editor of a Special Issues of *Physica Status Solidi A* and she is the guest editor of Special Issues of *Metals*, MDPI and *Frontiers in Materials*. She is the co-author of more than 100 papers in the materials science field and her current Citation Index is 21 (Scopus).

# Preface to "Metal Oxides"

Metal oxides represent a wide class of functional materials that exhibit a full spectrum of properties suitable for a large number of applications in many fields such as sensing, environmental remediation, energy storage and conversion, catalysis, optoelectronics, and photonics, to name only a few. Metal oxides' functional properties are strongly dependent on oxide's crystal structure, composition, native defects, doping, etc., which govern their optical, electrical, chemical and mechanical characteristics. Processing methods and growth parameters strongly determine the morpho-structural characteristics and therefore the physico-chemical properties of metal oxides. The band gap and electronic structure of oxides can be controlled and tailored by the size and dimension, resulting in a vast range of potential applications. The papers collected in this Special Issue include a miscellaneous composition encompassing several applications where metal oxides play a key role. Some papers also give insights into novel synthesis methods and processes aiming to reduce the negative environmental impacts and increase materials and process efficiency, thus also covering a broader concern on sustainability issues. As a guest editor of this Special Issue, I hope that the studies reported here will contribute to the advance of different research fields and will be of interest to the readers. I am grateful to all the contributing authors and to the editors involved in the creation of this Issue.

**Maria Luisa Grilli**
*Editor*

*Editorial*

# Metal Oxides

**Maria Luisa Grilli**

ENEA-Italian National Agency for New Technologies, Energy and Sustainable Economic Development, Energy Technologies Department, Casaccia Research Centre, Via Anguillarese 301, 00123 Rome, Italy; marialuisa.grilli@enea.it; Tel.: +390-630-486-234

Received: 15 June 2020; Accepted: 18 June 2020; Published: 19 June 2020

## 1. Introduction and Scope

Oxide materials in bulk and thin film form, and metal oxide nanostructures exhibit a great variety of functional properties which make them ideal for applications in solar cells, gas sensors, optoelectronic devices, passive optics, catalysis, corrosion protection, environmental protection, etc.

Metal oxide's functional properties are strongly dependent on oxide's crystal structure, composition, native defects, doping, etc., which govern their optical, electrical, chemical and mechanical characteristics. Processing methods and growth parameters strongly determine the morpho-structural characteristics and therefore the physico-chemical properties of metal oxides. The band gap and electronic structure of oxides can be controlled and tailored by the size and dimension, resulting in a wide range of potential applications.

This Issue is devoted to the modelling, synthesis and characterization of oxide thin films, multilayer structures (superlattices, metamaterials, devices, etc.) and nanomaterials with novel multifunctional characteristics which combine at least two excellent properties: electrical and optical, optical and mechanical, chemical and mechanical, thermal and chemical, etc.

Applications include: solar cells and optoelectronic devices; transparent conductive oxides (TCOs); plasmonics; photonic integrated circuits; chemical sensors; catalysis; corrosion protection; thermal protection; and energy conversion and storage.

## 2. Contributions

Ten articles have been published in the present Special Issue of *Metals*. Three of them are review papers.

The subjects are multidisciplinary, covering a wide range of applications, including (i) transparent conductive oxides (two papers); (ii) metal oxides composites and nanocomposites (two papers); (iv) welding and critical raw materials [1,2] (CRMs) saving (two papers); (v) metallurgical waste treatment (one paper); (vi) oxides for high temperature applications (thermal barrier coatings) (one paper); and (vii) nanostructured oxides and composites for gas sensing and desulfuration (one paper) and $CO_2$ capture (one paper).

The review paper from D'Anna et al. [3] deals with the notable synthesis of $WO_3$ (films and nanostructures) in ionic liquids (ILs). The synergy between ILs and metal oxides has been proposed recently to both direct oxides' production towards controllable nanostructures (nanorods, nanospheres, core-shell nanostructures, etc.) and to modify the metal oxide structure (incorporating ILs) in order to increase the gas adsorption ability, and thus the catalytic efficiency. Synergy between ILs and metal oxides can make a considerable contribution to the field of air pollutant sensing and remediation.

The review from Motoc et al. [4] deals with a novel and sustainable approach for doping zirconium oxide with mixed rare earth oxides (REOs) in the natural composition as extracted from monazite mineral. This allows to reduce the long and complex processes needed for the extraction, separation and purification of single rare earths, reducing greatly the costs and the environmental impact. Preliminary results are reported showing the ability of the mixed REOs, as occurring in the natural composition, to be an efficient zirconia dopant for thermal barrier coatings.

The review from Yilmaz et al. [5] reports on the bibliometric analysis of publications about In-doped ZnO, to reveal the general research tendency in the study of this transparent conductive oxide. Bibliometrics is an emerging cross-disciplinary discipline based on statistic and mathematic tools to map the state of the art and the development in a given area of scientific knowledge. This study can be a guide for researchers involved in the development of In-doped ZnO films and nanostructures for optoelectronics, solar cells and gas sensors applications.

The paper from Seok et al. reports on the characteristics of ITO films sputtered on flexible invar metal foils to be used as transparent electrodes substrates for curved perovskite solar cells (PSCs). Preliminary results indicate that invar metal foils are promising flexible substrates to substitute typical flexible polymer substrates for high-performance curved PSCs [6].

Two of the papers from Balos et al. [7,8] deal with the study of the influence of metallic oxide nano- and submicron particles for the performance of the activated tungsten inert gas (A-TIG) welding of austenitic stainless steels. Oxide coatings may have an interesting role in welding technology as catalysts of the TIG welding process. The method may help in saving CRMs because the consumables used in the welding of austenitic stainless steels contain critical raw materials (CRMs), or nearly CRMs and relatively expensive materials such as chromium, nickel and silicon metal.

The topic studied in Pan et al.'s paper [9] is the direct reduction of the ironmaking process of a rotary hearth furnace (RHF) as an effective method for the treatment of metallurgical wastes. A new RHF process was proposed to avoid the generation of sediments and to maximize the use of waste from the metallurgical process, thus improving the RHF energy efficiency.

The effect of various spark plasma sintering (SPS) temperatures on the properties of TiC- and TiN-reinforced alumina–zirconia composites for the precision machining of hard-working pieces was investigated in the paper by Szutkowska et al. [10]. Results demonstrate that upon increasing the sintering temperature, improvement in wear resistance and an increase in fracture toughness are observed in the tested samples. Properties of composites sintered in the case of pressure-assisted SPS were significantly better than those obtained by pressureless sintering (PS) at higher temperatures.

Another paper by Balos et al. [11] aims to study the influence of $TiO_2$ nanoparticles on the wear resistance of a Co-based hard-facing electrode. The hard-faced layer was produced using the common shielded metal arc welding (SMAW) technique. Results indicate that the wear resistance and hardness values of the hard-faced layers obtained with the $TiO_2$ nanoparticle coated on the SMAW electrode are higher with respect to the layers obtained with untreated electrodes.

Finally, the last published paper from Luisetto et al. [12] reports on the study of $CaO$-$CaZrO_3$ sorbents synthesized using the self-combustion method. $CaZrO_3$ was introduced into $CaO$-based sorbents to increase stability during repeated $CO_2$ capture/release cycles. The best stability was attributed to the correct balance between $CaO$, the active component, and the $CaZrO_3$ nanoparticles. The experimental data corroborated the adoption of the shrinking core spherical model for the interpretation of $CaO$ conversion to $CaCO_3$.

### 3. Conclusions and Outlook

Papers collected in this Special Issue compose a miscellaneous encompassing several research topics where metal oxides play a fundamental role. Some papers give also insights into novel synthesis methods and processes which can be guides to researchers for future studies.

The studies covered here offer richness and substantial depth on various topics, also taking into account the concern to reduce the negative environmental impacts and increase materials and process efficiency, thus covering a broader concern on sustainability issues.

As guest editor of *Metal Oxides*, I hope that the papers of this Issue will catch the interest of many scientists, will be useful for their future work and contribute to advance the different research fields.

**Acknowledgments:** I am grateful to the authors who contributed to this Special Issue, to the many anonymous reviewers who reviewed the manuscripts and to editors of *Metals* for their support during the preparation of this volume. In particular, my sincere thanks go to Betty Jin, assistant editor, for her continuous support, patience and valuable assistance in the volume preparation.

**Conflicts of Interest:** The author declares no conflict of interests and no involvement in the handling and reviewing of the co-authored papers of this Special Issue.

## References

1. Communication from the Commission to the European Parliament and the Council. The Raw Materials Initiative—Meeting Our Critical Needs for Growth and Jobs in Europe. Available online: http://eurlex.europa.eu/LexUriServ/LexUriServ.do?uri=COM:2008:0699:FIN:en:PDF (accessed on 12 June 2020).
2. Study on the Review of the List of Critical Raw Materials, Criticality Assessments. Available online: http://hytechcycling.eu/wp-content/uploads/Study-on-the-review-of-the-list-of-Critical-Raw-Materials.pdf (accessed on 12 June 2020).
3. D'Anna, F.; Grilli, M.L.; Petrucci, R.; Feroci, M. $WO_3$ and Ionic Liquids: A Synergic Pair for Pollutant Gas Sensing and Desulfurization. *Metals* **2020**, *10*, 475. [CrossRef]
4. Motoc, A.M.; Valsan, S.; Slobozeanu, A.E.; Corban, M.; Valerini, D.; Prakasam, M.; Botan, M.; Dragut, V.; Vasile, B.S.; Surdu, A.V.; et al. Design, Fabrication, and Characterization of New Materials Based on Zirconia Doped With Mixed Rare Earth Oxides: Review and First Experimental Results. *Metals* **2020**, *10*, 746. [CrossRef]
5. Yilmaz, M.; Grilli, M.L.; Turgut, G. A Bibliometric Analysis of the Publications on In Doped ZnO to be a Guide for Future Studies. *Metals* **2020**, *10*, 598. [CrossRef]
6. Seok, H.-J.; Kim, H.-K. Study of Sputtered ITO Films on Flexible Invar Metal Foils for Curved Perovskite Solar Cells. *Metals* **2019**, *9*, 120. [CrossRef]
7. Balos, S.; Dramicanin, M.; Janjatovic, P.; Zabunov, I.; Pilic, B.; Goel, S.; Szutkowska, M. Suppressing the Use of Critical Raw Materials in Joining of AISI 304 Stainless Steel Using Activated Tungsten Inert Gas Welding. *Metals* **2019**, *9*, 1187. [CrossRef]
8. Balos, S.; Dramicanin, M.; Janjatovic, P.; Zabunov, I.; Klobcar, D.; Busic, M.; Grilli, M.L. Metal Oxide Nanoparticle-Based Coating as a Catalyzer for A-TIG Welding: Critical Raw Material Perspective. *Metals* **2019**, *9*, 567. [CrossRef]
9. Pan, Y.; She, X.; Wang, J.; Liu, Y. Study of the Deposition Formation Mechanism in the Heat Exchanger System of RHF. *Metals* **2019**, *9*, 443. [CrossRef]
10. Szutkowska, M.; Cygan, S.; Podsiadło, M.; Laszkiewicz-Łukasik, J.; Cyboroń, J.; Kalinka, A. Properties of TiC and TiN Reinforced Alumina–Zirconia Composites Sintered with Spark Plasma Technique. *Metals* **2019**, *9*, 1220. [CrossRef]
11. Balos, S.; Janjatovic, P.; Dramicanin, M.; Labus Zlatanovic, D.; Pilic, B.; Hanus, P.; Jaworska, L. Microstructure, Microhardness, and Wear Properties of Cobalt Alloy Electrodes Coated with $TiO_2$ Nanoparticles. *Metals* **2019**, *9*, 1186. [CrossRef]
12. Luisetto, I.; Mancini, M.R.; Della Seta, L.; Chierchia, R.; Vanga, G.; Grilli, M.L.; Stendardo, S. $CaO–CaZrO_3$ Mixed Oxides Prepared by Auto-Combustion for High Temperature $CO_2$ Capture: The Effect of CaO Content on Cycle Stability. *Metals* **2020**, *10*, 750. [CrossRef]

 *metals*

*Review*

# WO$_3$ and Ionic Liquids: A Synergic Pair for Pollutant Gas Sensing and Desulfurization

**Francesca D'Anna [1],*, Maria Luisa Grilli [2],*, Rita Petrucci [3] and Marta Feroci [3],***

[1]  Dept. STEBICEF, University of Palermo, Viale delle Scienze, Build. 17, 90128 Palermo, Italy
[2]  Italian National Agency for New Technologies, Energy and Sustainable Economic Development (ENEA), Energy Technology Department, Casaccia Research Center, Via Anguillarese 301, 00123 Rome, Italy
[3]  Dept. Fundamental and Applied Sciences for Engineering (SBAI), Sapienza University of Rome, via Castro Laurenziano, 7, 00161 Rome, Italy; rita.petrucci@uniroma1.it
*  Correspondence: francesca.danna@unipa.it (F.D.A.); marialuisa.grilli@enea.it (M.L.G.); marta.feroci@uniroma1.it (M.F.); Tel.: +39-23-897540 (F.D.A.); +39-06-41733132 (M.L.G.); +39-06-49766563 (M.F.)

Received: 9 March 2020; Accepted: 2 April 2020; Published: 4 April 2020

**Abstract:** This review deals with the notable results obtained by the synergy between ionic liquids (ILs) and WO$_3$ in the field of pollutant gas sensing and sulfur removal pretreatment of fuels. Starting from the known characteristics of tungsten trioxide as catalytic material, many authors have proposed the use of ionic liquids in order to both direct WO$_3$ production towards controllable nanostructures (nanorods, nanospheres, etc.) and to modify the metal oxide structure (incorporating ILs) in order to increase the gas adsorption ability and, thus, the catalytic efficiency. Moreover, ionic liquids are able to highly disperse WO$_3$ in composites, thus enhancing the contact surface and the catalytic ability of WO$_3$ in both hydrodesulfurization (HDS) and oxidative desulfurization (ODS) of liquid fuels. In particular, the use of ILs in composite synthesis can direct the hydrogenation process (HDS) towards sulfur compounds rather than towards olefins, thus preserving the octane number of the fuel while highly reducing the sulfur content and, thus, the possibility of air pollution with sulfur oxides. A similar performance enhancement was obtained in ODS, where the high dispersion of WO$_3$ (due to the use of ILs during the synthesis) allows for noteworthy results at very low temperatures (50 °C).

**Keywords:** WO$_3$; ionic liquids; gas sensor; pollutant gases; desulfurization

---

## 1. Introduction

Tungsten trioxide (WO$_3$) is a n-type semiconductor widely investigated both in its doped and undoped forms, in powders, films and nanostructures, because of its good gas sensing, antibacterial and antimicrobial properties, its pH sensitivity, its photocatalytic activity for water-splitting, etc. A wide range of applications in several technological areas such as photocatalysis [1–3], gas sensing [4–8] and electrochromism [9–14] has been demonstrated. The reason for such wide applications lies in the semiconducting properties of WO$_3$, its polymorphous structure, its optical characteristics and its wide band gap. Tungsten trioxide shows several temperature dependent phase transitions: at room temperature, and up to 133 °C, the stable phase is the monoclinic one I ($\gamma$-WO$_3$). Upon heating above 330 °C, $\gamma$-WO$_3$ is converted to orthorhombic $\beta$-WO$_3$, which is stable up to 740 °C. At $T > 740$ °C the tetragonal $\alpha$-WO$_3$ phase is found. A metastable phase, the hexagonal WO$_3$ (h-WO$_3$) may also be obtained by opportune chemical synthesis [15,16] with potential advantages over the larger band gap $\gamma$-WO$_3$ phase [17]. In WO$_3$ powders, doped with H, Na, Li or other impurity atoms or in WO$_3$ thin film form, the cubic c-WO$_3$ phase also occurs [18,19]. This cubic phase is considered as the ideal high temperature phase and is consequently used as the reference for the structure of WO$_3$ [20]. The cubic perovskite-like structure is shown in Figure 1 and consists of corner sharing of regular octahedra with oxygen atoms at the corners and tungsten atoms at the center of each octahedron.

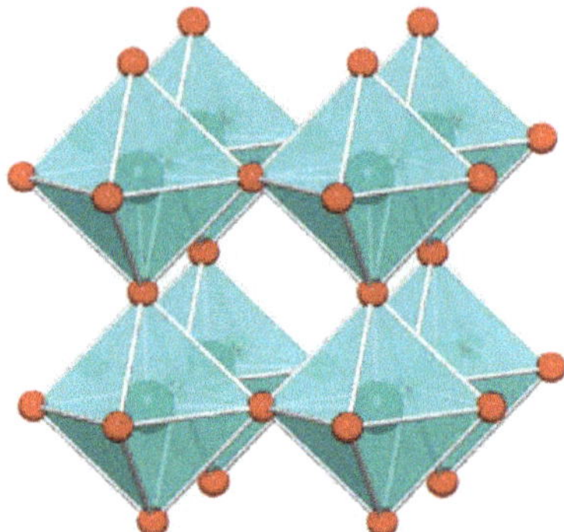

**Figure 1.** WO$_3$ cubic structure. Reproduced with permission [20]. Copyright 2019, Wiley.

The electronic bandgap value for WO$_3$ has been found in the range 2.6–3.3 eV, depending on the WO$_3$ phase and microstructure [9,21,22].

Together with SnO$_2$, WO$_3$ is the most widely used metal oxide semiconductor in commercial sensors due to its high sensitivity. Good sensing performances have been largely demonstrated for NO$x$ gas [23–27], but good detection of NH$_3$ [28,29], H$_2$S [30], H$_2$ [31] and SO$_2$ [32] was also demonstrated. In addition, due its good catalytic properties, WO$_3$ was also used as the metal oxide auxiliary phase in high temperature electrochemical sensors for NO$_2$ and CO detection [5] and for on board diagnostic (OBD) [33,34].

The gas sensing mechanism of metal oxide semiconductors is due to the resistivity changes in the presence of the adsorbed gas. According to Yamazoe and Shimanoe [35], the power laws that describe the change in semiconductors' resistance under exposure to a target gas can be derived by combining a depletion theory of semiconductors, which deals with the distribution of electrons between surface state (surface charge) and bulk, with the dynamics of adsorption and/or reactions of gases on the surface, which is responsible for the accumulation or reduction of surface charges. The role played by the negatively charged oxygen adsorbates on the sensing characteristic of the semiconductor gas sensors is the most extensively accepted explanation. In air the surface of a metal oxide is covered by several oxygen adsorbates such as O$_2{}^-$, O$^-$ and O$^{2-}$. In the case of n-type semiconductors, these oxygen adsorbates build a space-charge region on the surface of the metal oxide grains, resulting in an electron-depleted surface layer due to the electron transfer from the grain surface to the adsorbates. The depth of the space-charge layer is a function of the surface coverage of oxygen adsorbates and intrinsic electron concentration in the bulk.

The resistance of an n-type semiconductor gas sensor in air is rather high due to the development of a potential barrier to electronic conduction at each grain boundary. In the presence of a reducing gas, free charge carriers are released to the conduction band, whereas the reaction product desorbs thermally from the semiconductor. The electrons trapped by the oxygen adsorbates are transferred back to the oxide grains leading to a decrease in the potential barrier height and a resistance drop, as depicted in Figure 2 [36].

The resistance of a metal oxide semiconductor, and therefore its sensing properties, is affected by the depth of the space charge region L and by the crystallite size D of the material, as shown in Figure 3. The sensor element may be described as consisting of a chain of uniform crystallites of size D connected to each other by the necks of grain boundaries. Depending on the relative size of D and L, three different cases may occur [37]. If D is much greater than 2L (D >> 2 L), most of the volume of the crystallite is unaffected by the surface interaction, and resistivity is dominated by grain boundaries. When D > 2L, the grain size decrease in the depletion region extends deeper into the grains and the resistivity is controlled by the neck between grains.

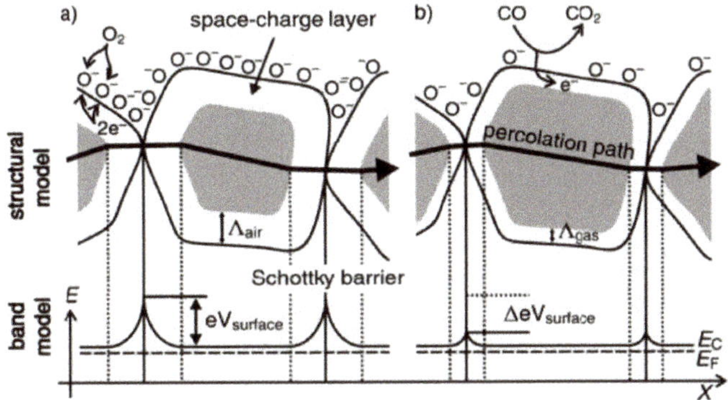



**Figure 2.** Structural and band models of conductive mechanism in n-type metal oxide semiconductor upon exposure to target gas, (**a**) with or (**b**) without CO [36].

When D < 2L the depletion region extends in the whole grain, the crystallites are almost fully depleted of electrons and resistance is dominated by the grain size effect. In this latter case of grain control sensing mechanism with a higher sensitivity and therefore a higher gas response are observed with respect to the other two cases, because of the larger quantity of adsorbates which can react with the target gas. The use of ultrafine grains of D comparable with or less than 2L is to achieve the state where the transducer function of the elements is operated by the grain-control mechanism [37].

It is therefore extremely important to reduce the calcination temperatures of the materials and/or the operating temperature of the gas sensor to avoid grain size increase. The morphology of the semiconducting oxide also plays a fundamental role in gas sensing performances. Each morphology has its own advantages, contributing to increased active sites on the surface, accelerated response speeds and enhanced gas diffusion. Among the various morphologies, hollow nanostructures and core-shell nanostructures show superior performances due to their larger specific surface areas, which allow both the inner and the outer surface to absorb the target gases [38].

The gas sensing of many $WO_3$ nanostructures has been investigated: nanoparticles, nanospheres [39,40], nanosheets [41], nanorods [42], nanowires [43,44]. In addition, doping with noble metals (Au, Ag, Pt and Pd) effectively enhances the catalytic properties of $WO_3$ and metal oxide semiconductors [45].

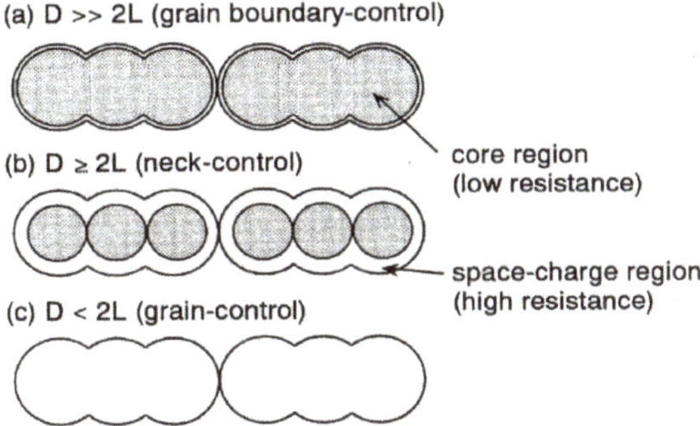



**Figure 3.** Schematic model of the grain size effect on a metal oxide semiconductor. Grain is represented by grey colour and depletion layer by white colour. Reproduced with permission [37]. Copyright 1991, Elsevier.

Several methods have been reported for the synthesis of $WO_3$ powders and nanostructures, such as thermal decomposition [46], precipitation [47], hydrothermal synthesis [48–50], electrodeposition [51], sol gel [52], etc.

Thin $WO_3$ films are usually grown using physical vapor deposition (PVD) techniques such as radio frequency sputtering [10,53], pulsed laser deposition [54], thermal [55] and e-beam evaporation, [56] or chemical vapor deposition (CVD) [57].

*Ionic Liquids*

Ionic liquids (ILs) have been known about since the late 1980s. They have been defined as salts having melting temperatures below 100 °C [58]. This general definition includes the more specific one related to room temperature ionic liquids (RTILs), i.e., salts liquid at room temperature. As a consequence of the above feature, RTILs can be used as solvents, and since their first appearance in literature, they have been claimed as an eco-friendly alternative to conventional organic solvents. ILs are formed by organic cations and organic or inorganic anions (Scheme 1).

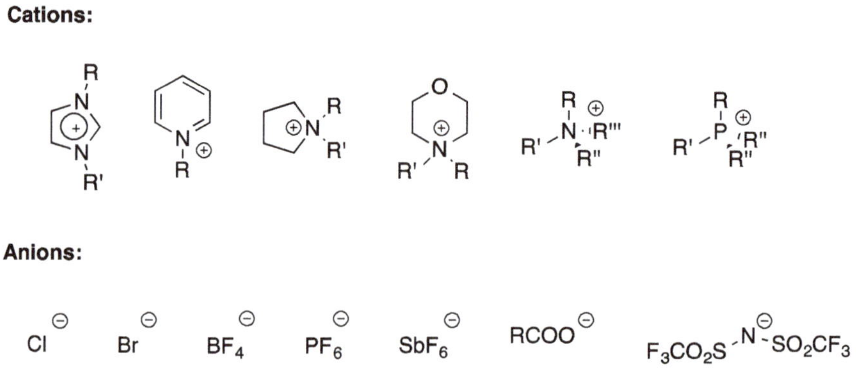

**Cations:**

**Anions:**

**Scheme 1.** Structures of most common cations and anions.

This endows them with low vapor pressure and flammability and high thermal stability. Consequently, their use allows all environmental issues generally deriving from the volatility of organic solvents to be avoided. The above-mentioned properties heavily depend on the cation or anion structure. On this subject, the melting point is determined by ion symmetry, as well as by the length of the alkyl chain on the cation. Properties like density and viscosity are affected, beyond the alkyl chain length, also by the cation hydrogen bond donor or the anion coordination ability. The latter feature also influences ILs thermal stability [59,60].

ILs also show high solubilizing ability and, being formed only by ions, high conductivity. The first feature explains why ILs have been applied in different fields of chemistry research. Indeed, they are able to dissolve both organic and inorganic small solutes, but also polymeric materials. In the latter case, they proved very efficient in dissolving natural polymers hard to solubilize in water or organic solvents, like cellulose and lignin [61,62].

Thanks to their conductive behavior they have been widely used in electrochemistry, in the preparation of lithium batteries, but also as electrolytes in dye sensitized solar cells. In some of the above-mentioned applications, IL use is advantageous with respect to inorganic electrolytes, as a consequence of their lower corrosivity [63,64].

Notwithstanding the plethora of different applications, the use of ILs as reaction media is probably the most widely applied and investigated [65–70]. They have been used to perform classical organic reactions, such as nucleophilic aromatic substitution, elimination reaction, ring to ring interconversion in heterocyclic systems, cycloaddition reaction and so on [71–77].

Interestingly, ILs, and in particular imidazolium-based ILs, proved very efficient in performing organocatalyzed reactions, taking advantage from the high activity of electrogenerated *N*-herocyclic carbenes [78–83].

As far as inorganic materials are concerned, ILs also interact well with metallic species and, in this context, their use in combination with metal oxides, with the aim to obtain efficient catalytic systems or sensors, is a very active area of research.

To this aim, they have been used to obtain gas sensors [84], but also to prepare noble metal clusters to be used in alcohol oxidation processes [85]. Independently from the nature of the processes in which they are used, ILs are frequently able to improve performance of processes. Indeed, they are able to increase both yield and reaction rates, but in some cases, they are also able to decrease the temperature of the processes. All the above effects allow their classification as valuable alternative to conventional solvents to be justified.

All of the above advantageous effects are generally explained taking in consideration the ionic nature of these solvents and considering the possibility to tailor properties of ILs to the features of the performed processes. Indeed, the behavior of ILs can be significantly changed, bringing small variations to cation or anion structures.

On this subject, it is worth noting that two different points of views are frequently detected in literature about the effect of ILs, as reaction media. The first one considers ILs as salt solutions and explains the effect using classical solvent parameters [86–88]. Differently, in some other cases, above all in the presence of aromatic ILs, the effects are rationalized considering the supramolecular network that features these solvents, justifying their description as polymeric supramolecular fluids [89,90].

In addition to simple ILs, task specific ionic liquids (TSILs) must also be considered. In this case a catalytic function borne on the cation or anion structure endows the salt not only with solvent but also catalyst function [85,91,92].

Irrespective of the nature and function of the IL, one of the most important advantages in using ILs lies in the possibility of their reuse, as simple liquid-liquid extraction allows the solvent/catalyst to be obtained in its pure form, ready for recycling. Clearly, the above aspect is relevant not only from an economical point of view, but above all from an environmental point of view. Indeed, in many cases they can be reused for at least five cycles, inducing a small decline in the degree of solution, as observed in the phosphorylation of corn starch using 1-butyl-3-methylimidazolium chloride as solvent medium [93].

Other systems that allow efficient IL recycling are the supported ionic liquid phases (SILPs) [94–96] or systems in which ILs are immobilized in a gelatinous network, giving rise to the obtainment of the so-called ionogels [97–101].

Besides the applications, the biological and environmental effects of these solvents have also been recently addressed to minimize the impact at the time of disposal. The tailoring property is very useful also in this context. Indeed, a suitable choice of substituents on the cation or anion structure allows the impact of these solvents to be significantly decreased, both on the environment and human health [102]. On this subject, aliphatic cations are generally preferred to aromatic ones, and among the anions, the ones deriving from amino acids, alkyl sulphates, halides and sugar-based anions are considered the most environmentally friendly [103].

Lastly, biobased ILs have also recently played a pivotal role. Indeed, the possibility of preparing these solvents using waste materials represents a way to improve their life cycle [104]. Indeed, several building blocks, including sugars, aminoacids, amino alcohols and so on, can be used as a precursor of ILs, and biobased ILs are solvents that should be suitable for processing their starting material. This, as recently reported by Socha et al., should allow a closed-loop biorefinery able to satisfy its own need of solvent to be realized [105].

As previously stated, ILs can be successfully used in various application fields. The synergic effect of $WO_3$ and ionic liquids in gas sensing and desulfurization reactions has been only recently studied, and this can be attributed to the ability of ILs to dissolve inorganic compounds (thus allowing the production of composites with a high active area) and to the solubility of water and permeability of some gases in certain ionic liquids, thus allowing the concentration of analytes (gases) on the surface of the sensor, while lowering the negative effect of high levels of humidity (vide infra).

We herein want to review the contextual and synergic use of $WO_3$ and ionic liquids in the field of air pollution, from the point of view of the quantification of gaseous air contaminants (pollutant gas sensing, paragraph 2) and from the point of view of the possibility of decreasing the extent of $SO_x$ liberation in the atmosphere due to impure fuel combustion (hydrodesulfurization and oxidative desulfurization, paragraph 3).

## 2. $WO_3$ and Ionic Liquids in Pollutant Gas Sensing

Air quality assessment is one of the main tasks of recent (and future) years. In fact, "Air pollution kills an estimated seven million people worldwide every year. WHO data show that 9 out of 10 people breathe air containing high levels of pollutants. WHO is working with countries to monitor air pollution and improve air quality" [106]. In particular, air pollution can be divided into two different categories (based on ambient and pollution source): indoor and outdoor. Indoor (household) air pollution is mainly due to incorrect fuel utilization (in cooking or stoves), while outdoor (ambient) air pollution has a larger number of sources [107]:

- Fuel combustion from motor vehicles (e.g., cars and heavy-duty vehicles),
- Heat and power generation (e.g., oil and coal power plants and boilers),
- Industrial facilities (e.g., manufacturing factories, mines and oil refineries),
- Municipal and agricultural waste sites and waste incineration/burning,
- Residential cooking, heating and lighting with polluting fuels.

It is thus evident that cheap, light, easy-to-use, non-polluting and/or recyclable gas sensors are the topic of current research [108,109]. One of the main differences between indoor and outdoor gas sensing is the very different level of humidity of these two ambients, very high outdoor and controllable indoor. High levels of humidity can compromise the operations of the sensor devices [110].

Many are the materials used in air pollutant sensors, and among them semiconducting metal oxides are very popular [111]. In particular, $WO_3$, due to its peculiarities, is often used as sensor material [112–115].

More recently, in order to enhance the performance of $WO_3$ sensors, ILs have been used both in the $WO_3$ production process (i.e., incorporating ILs in $WO_3$ structures or influencing such a structure) and in the sensor device construction process (i.e., incorporating ILs in the device). In both cases better performances were obtained in comparison to the same sensors obtained without using ILs.

Li and coworkers reported the use of two very common imidazolium ILs (1-butyl-3-methylimidazolium and 1-carboxymethyl-3-methylimidazolium chlorides, BMImCl and CMImCl, respectively) in the synthesis of nanostructured $WO_3$ particles [116]. In particular, nanorods could be selectively obtained using 1-carboxymethyl-3-methylimidazolium chloride, while nanoparticle-constructed spheres could be selectively obtained using 1-butyl-3-methylimidazolium chloride, in both cases starting from $WCl_6$.

The effect of the nature and amount of imidazolium ionic liquids on the morphology of the produced nanoparticles was evidenced, and a hypothesis of mechanism was reported. In all cases, starting from $WCl_6$ as tungsten source and after calcination, monoclinic $WO_3$ was obtained, in pure form and well crystallized, but in the absence of IL, nanoplates were obtained (1 hundred nm size), while in the presence of CMImCl nanorods (30 nm diameter), and in the presence of BMImCl, spheres (1500 nm size, formed by subunits of 40 nm size) were obtained. The morphology of produced nanostructured $WO_3$ not only depends on the nature of the substituents on the imidazolium cation (butyl or carboxymethyl), but also on the amount of IL. In fact, in the case of CMImCl, only an equimolar amount of IL (with respect to $WCl_6$) led to the formation of nanorods, while increasing the IL amount led to irregular particles being formed. Using BMImCl (butyl group on the imidazolium cation), the situation was the opposite: an equimolar amount of IL led to the formation of irregular particles, while an excess of IL yielded nanoparticle-constructed spheres.

The authors suggested that in the case of CMImCl, the IL can be selectively adsorbed on the surface of crystals, thus leading to an alteration of the anisotropic growth process. In the case of BMImCl, the hypothesis is that this IL can act as a template for the growth of nanospheres. The spheres, once formed, aggregate by different kind of interactions (electrostatic, hydrogen bonding, π-π stacking), yielding the nanoparticle-constructed spheres.

The produced $WO_3$ nanoparticles were used in conductimetric sensors for the detection of various pollutant gases (ethanol, methanol, isopropanol, ethyl acetate, toluene); the optimal temperature of the sensors was 240–300 °C, with short response (3–25 s, depending on the analyte and its concentration) and recovery times, and with 5 ppm as the lower detection limit.

Furthermore, Cheng and coworkers reported the use of an imidazolium IL (1-butyl-3-methylimidazolium tetrafluoroborate) in the process of $WO_3$ nanoparticle formation [117]. The nanoparticles were characterized using IR, XPS and XRD analyses. The produced nanoparticles were tested in a $NO_2$ electrochemical sensor, showing enhanced properties with respect to the corresponding $WO_3$ nanoparticles sensor in which no IL was used. The authors suggested that the better performances could be due to residue IL on the surface of the particles, with increased oxygen adsorption. The lower detection limit was 0.1 ppm, with a very high selectivity towards $NO_2$ (in the presence of ethanol, nitrobenzene, acetone, ammonia, carbon dioxide, etc) at the noteworthy low temperature of 50 °C. The reaction is the adsorption and reduction of $NO_2$ (outlined in Figure 4), which, in the presence of oxygen, yields $NO_3^-$, followed by the formation of a superoxide anion. This process traps electrons from $WO_3$ particles, leading to an increase of sensor resistance.

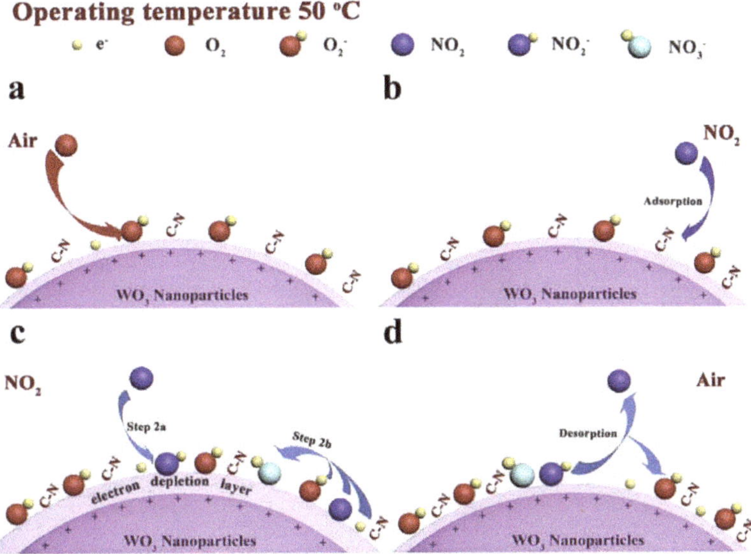


**Figure 4.** Proposed $NO_2$ detection mechanism. Reproduced with permission [117]. Copyright 2019, Elsevier.

As previously stated, outdoor gas detection suffers from high humidity levels, which in most cases lower the sensor performances. ILs can overcome (in part) this problem, as reported by Bendahan and coworkers, regarding a $WO_3$ sensor for aromatic volatile compounds (benzene, toluene, ethylbenzene and xylenes) [118]. A cheap and reliable removable filter, based on common imidazolium ILs, was englobed into the sensor, with no loss in sensitivity and selectivity, for the efficient removal of most of the humidity, with significant improvement of the sensor performances, as reported in Figure 5.

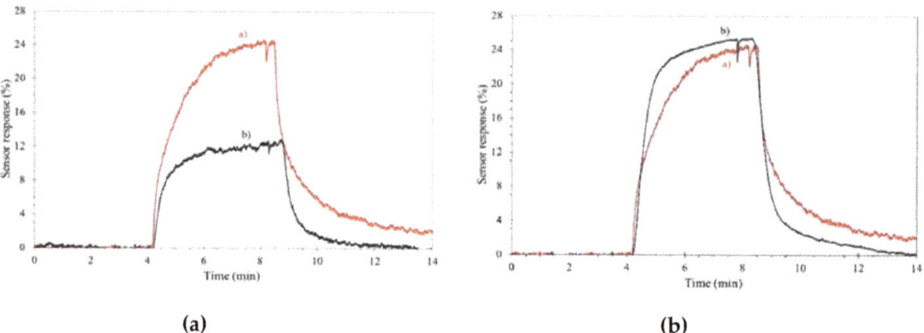

**Figure 5.** Effect of an ionic liquid (IL)-composed filter on aromatic volatile compound detection. Sensor response under 500 ppb compounds. (**a**): dry air; (**b**): wet (50%) air. Left, without filter, right with filter. Reproduced with permission [118]. Copyright 2018, ISFA.

Mahmoud and coworkers reported the use of $WO_3$ nanoparticles, obtained using a sol-gel method and glycerol IL, as components of a flexible poly-vinyl alcohol (PVA) membrane inserted between the two electrodes of a portable $H_2S$ electrochemical sensor [119]. Along with excellent reproducibility of the results and long-term stability, the best operation temperature was quite low (80 °C), with a reasonable response time (19.1 ± 3.4 s). This result is quite significative, as the majority of metal oxide-based sensors operate at higher temperatures (200 °C or higher). The detection limit of this sensor was 10 ppm of $H_2S$, also in the presence of noteworthy amounts of interferents ($H_2$, $C_2H_4$), as reported in Figure 6.

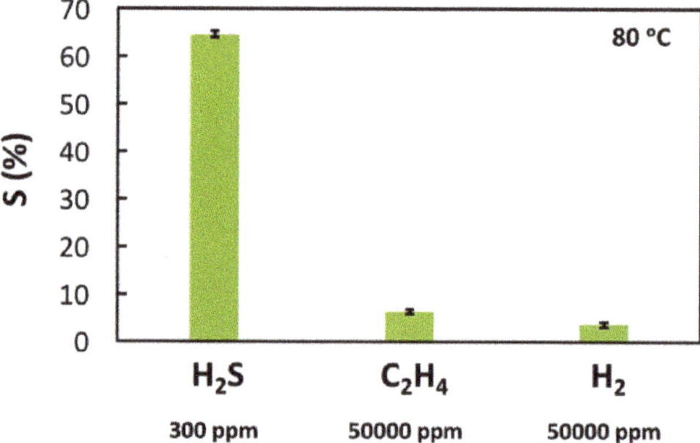

**Figure 6.** Selectivity of a $WO_3$ nanoparticles-glycerol IL-based $H_2S$ sensor when exposed to $H_2S$, $H_2$ and $C_2H_4$ at 80 °C. Reproduced with permission [119]. Copyright 2017, Elsevier.

This electrochemical sensor is based on electrochemical reduction of $H_2S$ in the presence of adsorbed oxygen, generating $SO_2$. The same authors reported an improvement of the $H_2S$ sensor using chitosan instead of poly-vinyl alcohol (PVA) in the flexible membrane (containing $WO_3$ nanoparticles and IL), obtaining an electrochemical sensor less humidity dependent and working at the very low temperature of 40 °C [84]. In this case, the cooperation of $WO_3$ nanoparticles, adsorbed oxygen, IL and chitosan seems to allow better results, as depicted in Figure 7.

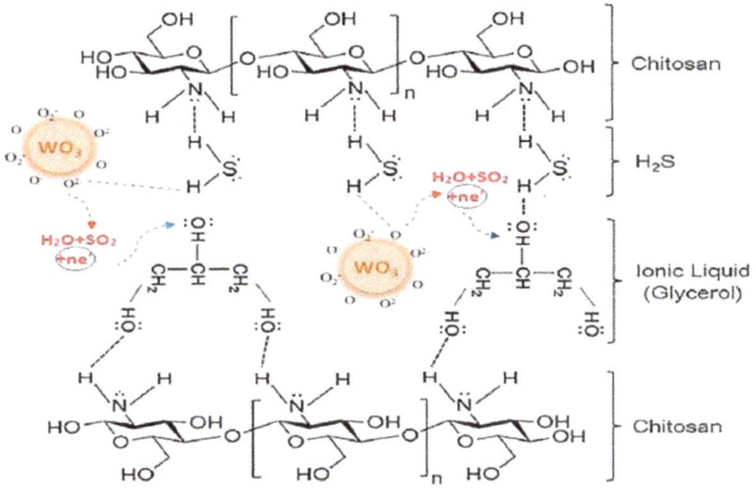

**Figure 7.** Participation of all sensor components in H$_2$S detection at 40 °C. Reproduced with permission [84]. Copyright 2020, Elsevier.

The possibility of using a common imidazolium IL (EMIm-BF$_4$) in a conductimetric CO$_2$ gas sensor was recently reported by Daves and Ersoez [120]. In particular, the authors exploited the properties of organic IL, while combining them with the mechanical stability of inorganic compounds. The electrochemical sensor was formed using an ionogel (obtained by the combination of the IL and an inorganic porous host) deposited over WO$_3$. The very good performances of this kind of sensor are due to the permeability to CO$_2$ of the ionogel layer (also compared to pure IL, due to the high enhancement of the surface for the presence of the porous inorganic host).

Although ILs have only recently started to be used to enhance the performances of metal oxide-based electrochemical gas sensors (with reference, in particular, to WO$_3$-based sensors), the recognition of their role has been clearly asserted. In fact, not only ILs do heavily influence the structure of WO$_3$ nanoparticles when used in the synthetic step (nanoplates, nanorods, nanospheres, etc.), but their residues, eventually present on the surface of nanoparticles, can improve the sensed gas adsorption, thus yielding better performances. We hope that the virtually infinite possibility to vary the structures of ILs (and thus their physico-chemical characteristics) will lead to the fabrication of organic-inorganic hybrid devices with optimal gas sensing performances, in terms of sensitivity, selectivity, reproducibility and low humidity dependence.

## 3. WO$_3$ and Ionic Liquids in Fuel Desulfurization

One of the main problems in fuel combustion (beside CO$_2$ formation) is the production of polluting and toxic gases, such as SO$_2$ and SO$_3$. In particular, naphtha shows the highest sulfur and olefine content, with a high octane number. The target of the desulfurization process is thus the elimination of the highest possible sulfur amount, while preserving a high olefine content (maintaining a high octane rating), in order to have good fuel performances during combustion, lowering to a minimum level the emissions of sulfurated compounds.

Of the organic sulfur compounds present in petroleum derivatives, thiols, sulfides, disulfides and thiophenes are among the most important (Figure 8).

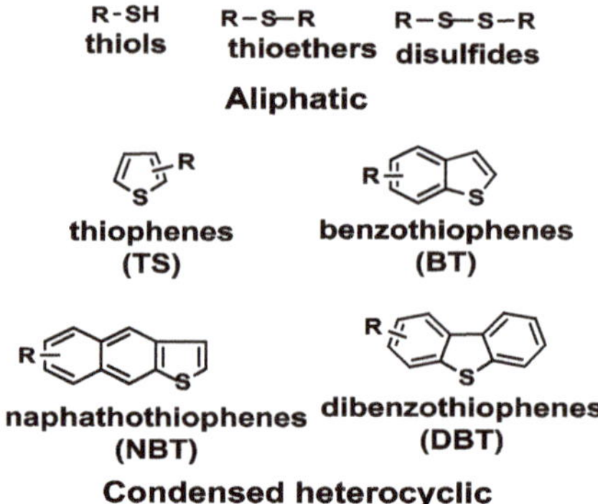



**Figure 8.** Main sulfur-containing compounds in petroleum fuels. Reproduced with permission [121]. Copyright 2016, Elsevier.

Regarding the available methods to carry out liquid fuel desulfurization, hydrodesulfurization (HDS) and oxidative desulfurization (ODS) are among the most used.

### 3.1. WO$_3$ and Ionic Liquids in Hydrodesulfurization (HDS)

Hydrodesulfurization (HDS) is the removal of sulfur compounds from liquid petroleum derivatives by reaction with molecular hydrogen at a high temperature (300–350 °C). High pressures of H$_2$ are needed for this reaction (20–100 atm), and sulfur compounds are transformed into H$_2$S and hydrocarbons in the presence of catalysts. H$_2$S is at the end removed from the reaction mixture and transformed into elemental sulfur [121]. This process is quite efficient to remove aliphatic compounds (thiols, sulfides, disulfides), while thiophene derivatives are quite recalcitrant, needing more drastic conditions, which influence the cost of the process, among other things. The difficulty of removing thiophenes seems to be related to a lower adsorption on the catalyst surface.

ILs can play an important role in catalyst synthesis, in order to obtain compounds with improved performances. To this regard, it should be remembered that HDS catalysts promote both sulfur transformation into H$_2$S and olefine saturation, which is a highly undesired side reaction (lowering the octane number of the produced fuel). Daage and Chianelli [122] and Topsøe [123] elaborated a model to understand the selectivity of desulfuration over olefine saturation using supported metal sulfide catalysts (the brim-edge model). In particular, the brim sites of a multistack metal sulfide HDS catalyst (the top and bottom layers) catalyze both hydrodesulfuration and olefine saturation, while the edge sites catalyze only hydrodesulfuration. The conclusion is that a good HDS catalyst contains a high edge to brim ratio, and this ratio can be in part controlled by controlling the interactions between the catalyst and its support (necessary to disperse the metal catalyst, enhancing its surface). Such an interaction should be a compromise in order to have good metal dispersion, minimizing side reactions. Moreover, the incorporation of a "promoting element" (usually Co or Ni) can increase the number of edge sites.

Bao, Yuan and coworkers reported that the use of an IL, tetraethylammonium bromide (TEAB), in an aqueous solution at room temperature allowed an organic-inorganic nanocomposite (TEA$_2$W$_6$O$_{19}$) to be formed [124]. The characteristic of such a composite was its core-shell structure (W$_6$O$_{19}$$^=$ core and TEA$^+$ shell), which led to monodispersion, useful to deposit it onto alumina (the support) in the presence of a Ni promoter. The amount of dispersed Ni, along with the kind of dispersion, created catalyst structures with different edge to brim ratios, as evidenced in Figure 9. For a useful comparison, a catalyst obtained using the conventional impregnation method is also reported (Figure 9d).

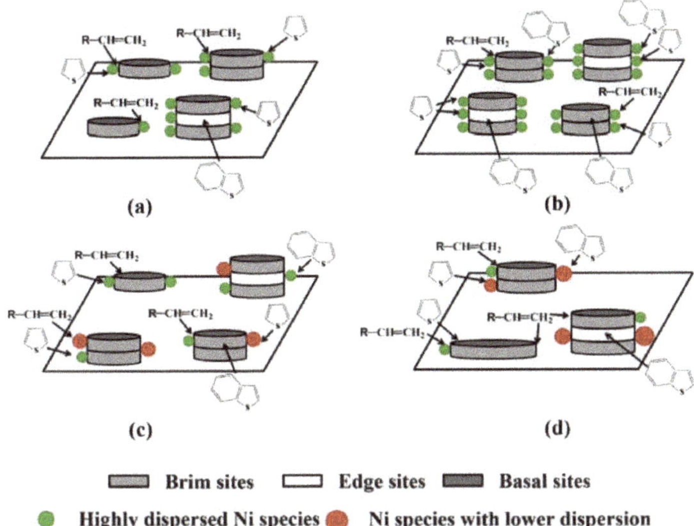



**Figure 9.** Schematic representation of different supported Ni promoted $WO_3/Al_2O_3$ catalysts. NiO content: (**a**) < 4.8 wt.%, (**b**) 4.8 wt.%, (**c**) > 4.8 wt.%, (**d**) 4.8 wt.% (impregnation method). Reproduced with permission [124]. Copyright 2015, Elsevier.

The catalysts thus obtained, with a compromise between metal dispersion and stacking, yielded enhanced performances, of which the improvement of HDS selectivity, minimizing olefine saturation, was of noteworthy importance. The ability of ILs to induce the formation of particular structures (in this case a core-shell one) allowed the improvement of the catalyst performances.

*3.2. WO$_3$ and Ionic Liquids in Oxidative Desulfurization (ODS)*

As previously said, thiophene derivatives are less prone to HDS reaction. In order to improve thiophenes abatement, while maintaining acceptable process costs, oxidative desulfurization (ODS) can be considered a promising method, due to its simplicity and high efficiency [121,125,126]. ODS is the chemical oxidation of sulfur compounds in liquid fuels (using as an example $H_2O_2$ as oxidant), yielding products which can be easily removed from the reaction mixture using a non-miscible solvent. In this regard, ILs can be efficiently used as extractive solvents of both starting and oxidized sulfur compounds and can be considered "greener" alternatives to conventional volatile organic compounds (VOCs).

As an example, Li and coworkers efficiently carried out the oxidative desulfurization of fuel using $H_2O_2$ as the oxidant agent in the presence of a $WO_3/C$ composite catalyst [127]. The extraction of sulfur compounds was carried out using an imidazolium IL (1-ethyl-3-methylimidazolium ethyl sulfate), added to the fuel as a non-miscible solvent (biphasic reaction medium). The $WO_3/C$ composite was oxidized to the complex $H_2[W_2O_3(O_2)_4(H_2O)_2]_2$ in the IL phase, which also extracted from fuel the aromatic sulfur compounds; the complex then oxidized dibenzothiophene (DBT) to its sulfone ($DBTO_2$), which remained in the IL phase, allowing an easy separation. The same process could be carried out using 1-butyl-3-methylimidazolium tetrafluoroborate as $DBTO_2$ extraction solvent [128].

In addition, Zhu, Li and coworkers reported the ability of an imidazolium IL ($C_{16}$MImBr) to direct the synthesis of a $WO_3$-$SiO_2$ composite towards a mesoporous material (W-$SiO_2$-20, Figure 10), which exhibited a high dispersion of tungsten throughout the structure (enhancing the catalytic activity) [129]. The synthesis of the mesoporous catalyst was carried out starting from a polyoxometalate compound ($[C_{16}mim]_3PW_{12}O_{40}$) in a one-pot gel of tetraethyl orthosilicate, which was then calcinated at 550 °C.

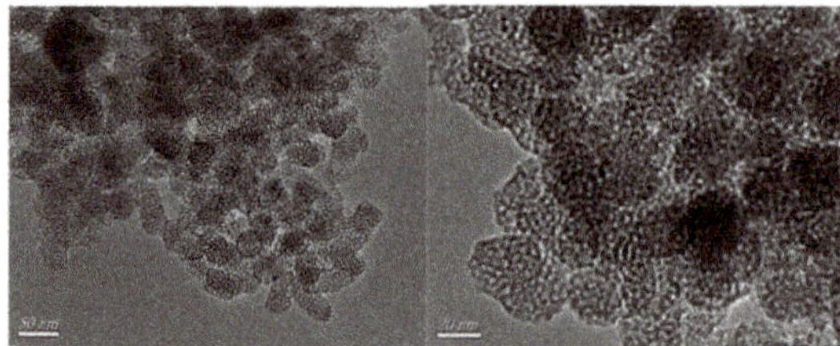

**Figure 10.** TEM images of the W-SiO$_2$-20 catalyst. Reproduced with permission [129]. Copyright 2016, Elsevier.

The mesoporous catalyst was characterized using the usual techniques and efficiently used in ODS reactions (Figure 11) at the low temperature of 60 °C. The reaction times were quite short and very good yields were obtained after only 30 min. Moreover, the process did not require additional organic solvents as extractants.

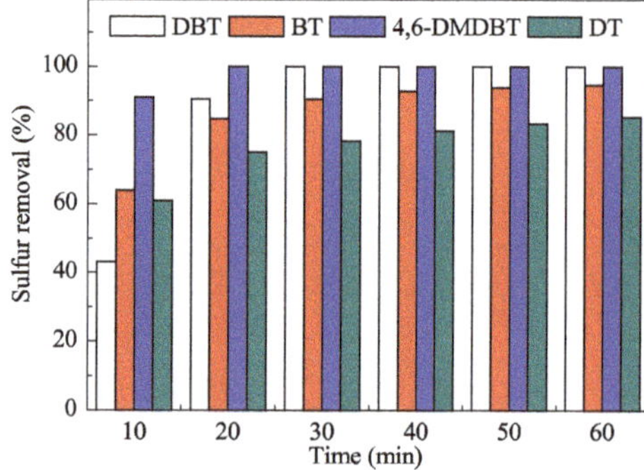



**Figure 11.** Removal of sulfur compounds in oxidative desulfurization (ODS) with the W-SiO$_2$-20 catalyst. DBT: dibenzothiophene; BT: benzothiophene; 4,6-DMDBT: 4,6-dimethyldibenzothiophene; DT: 1-dodecanethiol. Reproduced with permission [129]. Copyrights 2016, Elsevier.

The same authors reported a similar synthesis, utilizing a different support (in this case mesoporous ZrO$_2$) evidencing as both IL and calcination temperature, influenced the morphology and the dispersion of WO$_3$ [130]. The best obtained catalyst (calcinated at 700 °C, using a C$_{16}$-ammonium IL, 700-C$_{16}$-WO$_3$/ZrO$_2$) performed very well in oxidation desulfurization. Dibenzothiophene (DBT) could be completely oxidized to DBT sulfone (DBTO2). Moreover, the catalyst could be recycled ten times with very low efficiency loss.

A functional IL ([(C$_{16}$H$_{33}$)$_2$N(CH$_3$)$_2$]$_2$W$_2$O$_{11}$) acted as WO$_3$ nanoparticle precursor and a large surface area (203 m$^2$/g) few-layer g-C$_3$N$_4$ support was used to disperse them, yielding a supported catalyst [131]; this composite was characterized using SEM (Figure 12), TEM, FT-IR, XRD and XPS, showing highly dispersed nanoparticles. The analysis showed that during the synthetic process of WO$_3$ dispersion on the support, the structure of few-layer g-C$_3$N$_4$ was not destroyed (Figure 12D vs C), leading to a WO$_3$ catalyst with a very high surface, not obtainable using pure WO$_3$, whose structure showed agglomerates (Figure 12A).

**Figure 12.** SEM images of (**A**) WO$_3$, (**B**) bulk g-C$_3$N$_4$, (**C**) few-layer g-C$_3$N$_4$, (**D**) WO$_3$/few-layer g-C$_3$N$_4$. Reproduced with permission [131]. Copyright 2019, Elsevier.

The enormous amount of exposed active sites rendered the composite an excellent catalyst in ODS processes, with the removal of 100% refractory sulfur-containing molecules at 50 °C in 1 h. Moreover, the catalyst was recycled up to six times without efficiency loss. A possible reaction mechanism is depicted in Figure 13.

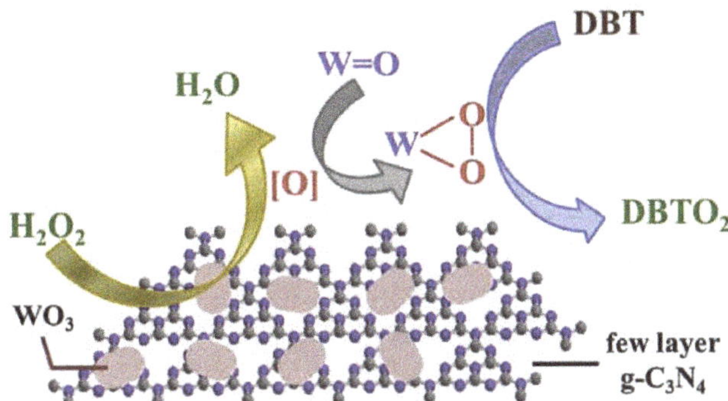


**Figure 13.** Possible ODS mechanism using WO$_3$/few-layer g-C$_3$N$_4$ as catalyst. Reproduced with permission [131]. Copyright 2019, Elsevier.

Last, tungsten trioxide-carbon nanotubes composite (WO$_3$/CNT) was demonstrated to be a very good catalyst in ODS of recalcitrant aromatic sulfur compounds, as reported by Li and coworkers [132]. The catalyst synthesis was quite easy and was carried out in the presence of an imidazolium IL (C$_{16}$MImCl), using phosphotungstic acid (HPW) as tungsten source (Figure 14).

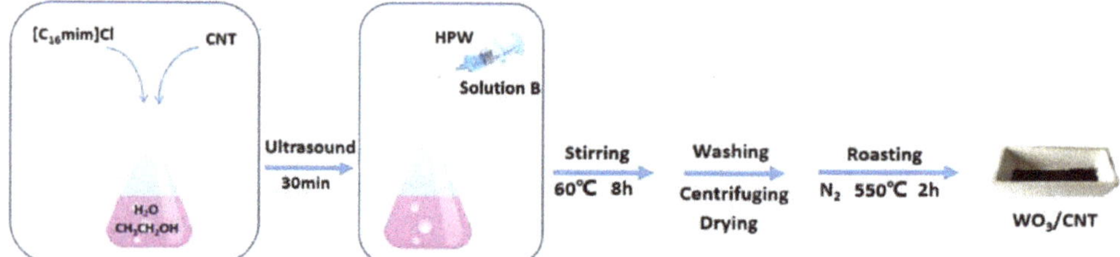


**Figure 14.** Schematic description of WO₃/CNT catalyst preparation. Reproduced with permission [132]. Copyright 2019, Wiley and Society of Chemical Industry.

The characterization of such a composite showed that the IL played a crucial role in determining the dispersion degree and the crystal phase of WO₃ on the carrier (carbon nanotube, CNT). In fact, the IL improved the transformation of tungsten trioxide from monoclinic to tetragonal, inhibiting at the same time the growth of metal oxide grains. In this way, high WO₃ dispersion was obtained, enhancing the catalytic activity. Moreover, a comparison of catalytic activity of different supported tungsten oxide forms was carried out, demonstrating the following activity order in sulfur oxidative removal: tetrahedral > tetragonal > monoclinic.

In conclusion, this section demonstrated the possibility of using an IL to enhance WO₃ dispersion on an inorganic support, ensuring the formation of composites with a very large number of exposed catalytic sites. In addition, in some cases the IL was demonstrated to be able to direct the morphology of WO₃ particles.

## 4. Conclusions and Perspectives

The synergy between WO₃ and ILs in increasing the performances of the corresponding devices is quite recent. During the last decade, ILs were demonstrated to be able to enhance the ability of tungsten trioxide in polluting gas sensing and in desulfurization processes, and the increasing number of publications on this topic gives an idea of future developments. This is due to the virtually infinite possibility to vary the structures of ILs (and thus their physico-chemical characteristics) changing the nature of cation and anion.

As ILs interact well with metal oxides (in particular with WO₃), they can be easily used as alternatives to classical organic solvents, with the advantage of minimizing air pollution (due to their virtually null vapor pressure) and the actual possibility of their recycling. Moreover, their ability to dissolve inorganic compounds can play a noteworthy role in producing highly dispersed composites, enhancing their activity. Furthermore, their hygroscopy can be successfully exploited in producing gas sensors with increased resistance to high humidity levels (especially for outdoor devices, for which humidity can be a serious problem). When using an IL in WO₃ production processes, it is possible to induce the formation of particular structures (nanorods, nanospheres, etc.), depending on the nature of the IL. Moreover, in some cases these salts are inglobated into the metal oxide structure, enhancing the performances of the corresponding gas sensor by increasing the gas adsorption on the surface of the composite.

It is thus probable that in the near future these peculiarities of ILs will lead to devices with better sensing performances, also lowering the environmental impact by choosing biocompatible ILs.

Another approach to air pollutants, besides their sensing and removal, is related to the decrease of their production. In particular, when referring to outdoor air pollution, fuel combustion plays a giant part. It is thus important to remove any possible source of pollution before combustion, in order to lower the environmental impact. As sulfur oxides are among the toxic pollutants deriving from fuel combustion, fuel desulfurization is a priority for the oil industry.

In this context, $WO_3$ in combination with ILs plays a pivotal role. In fact, ILs enhance the extent of adsorption of sulfur derivatives on the surface of the $WO_3$ catalyst, enabling better conversions also in the case of thiophene derivatives, which are among the most refractory sulfur compounds in fuel.

Although it is not possible, at this stage, to define the optimal $WO_3$ nanostructure morphology for all the applications reported in this review, the strong suggestion evinced from literature data is the ability of ILs to induce the formation of core-shell nanostructures allowing the improvement of the catalyst performances. The higher the surface to volume ratio, the better the performances of the $WO_3$ catalyst, both in the case of pure tungsten oxide compounds (higher number of active sites), and in the case in which IL fragments are present on the catalyst surface (allowing for a higher number of functionalized sites). Moreover, nanoparticles of 10-50 nm size seem optimal for the applications here reported.

Due to the increasing work of the scientific community in producing task-specific ILs showing better performances, along with an environmentally benign fingerprint (from both synthetic and disposal point of view, using biomasses as starting materials), we are confident that the synergy between ILs and metal oxides will make a considerable contribution to the field of air pollutant sensing and remediation.

**Author Contributions:** All authors contributed equally to this manuscript. All authors have read and agreed to the published version of the manuscript.

**Funding:** This research was funded by the University of Palermo and by Sapienza University of Rome, grant number RM11916B462FA71F.

**Acknowledgments:** The authors want to thank Marco Di Pilato for technical support.

**Conflicts of Interest:** The authors declare no conflict of interest. The funders had no role in the design of the study; in the collection, analyses, or interpretation of data; in the writing of the manuscript, or in the decision to publish the results.

## References

1. Kwong, W.L.; Savvides, N.; Sorrell, C.C. Electrodeposited nanostructured $WO_3$ thin films for photoelectrochemical applications. *Electrochim. Acta* **2012**, *75*, 371–380. [CrossRef]
2. Yu, J.; Qi, L. Template-free fabrication of hierarchically flower-like tungsten trioxide assemblies with enhanced visible-light-driven photocatalytic activity. *J. Hazard. Mater.* **2009**, *169*, 221–227. [CrossRef] [PubMed]
3. Amano, F.; Ishinaga, E.; Yamakata, A. Effect of Particle Size on the Photocatalytic Activity of $WO_3$ Particles for Water Oxidation. *J. Phys. Chem. C* **2013**, *117*, 22584–22590. [CrossRef]
4. Long, H.; Zeng, W.; Zhang, H. Synthesis of $WO_3$ and its gas sensing: A review. *J. Mater. Sci.: Mater. Electron.* **2015**, *26*, 4698–4707. [CrossRef]
5. Grilli, M.L.; Chevallier, L.; Vona, M.L.D.; Licoccia, S.; Bartolomeo, E.D. Planar electrochemical sensors based on YSZ with $WO_3$ electrode prepared by different chemical routes. *Sensors Actuator. B Chem.* **2005**, *111*, 91–95. [CrossRef]
6. Lu, R.; Zhong, X.; Shang, S.; Wang, S.; Tang, M. Effects of sintering temperature on sensing properties of $WO_3$ and Ag-$WO_3$ electrode for $NO_2$ sensor. *R. Soc. Open Sci.* **2018**, *5*, 171691. [CrossRef] [PubMed]
7. Dutta, A.; Kaabbuathong, N.; Grilli, M.L.; Di Bartolomeo, E.; Traversa, E. Study of YSZ-Based Electrochemical Sensors with $WO_3$ Electrodes in $NO_2$ and CO Environments. *J. Electrochem. Soc.* **2003**, *150*, H33–H37. [CrossRef]
8. Yamazoe, N. New approaches for improving semiconductor gas sensors. *Sens. Actuators B Chem.* **1991**, *5*, 7–19. [CrossRef]
9. Granqvist, C.G. Electrochromic tungsten oxide films: Review of progress 1993–1998. *Sol. Energy Mater. Sol. Cells* **2000**, *60*, 201–262. [CrossRef]
10. Masetti, E.; Grilli, M.L.; Dautzenberg, G.; Macrelli, G.; Adamik, M. Analysis of the influence of the gas pressure during the deposition of electrochromic $WO_3$ films by reactive r.f. sputtering of W and $WO_3$ target. *Solar Energy Mater. Solar Cells* **1999**, *56*, 259–269. [CrossRef]
11. Rao, M.C. Structure and properties of $WO_3$ thin films for electrochromic device application. *J. Non-Oxide Glasses* **2013**, *5*, 1–8.

12. Arvizu, M.A.; Qu, H.-Y.; Cindemir, U.; Qiu, Z.; Rojas-González, E.A.; Primetzhofer, D.; Granqvist, C.G.; Österlund, L.; Niklasson, G.A. Electrochromic $WO_3$ thin films attain unprecedented durability by potentiostatic pretreatment. *J. Mater. Chem. A* **2019**, *7*, 2908–2918. [CrossRef] [PubMed]

13. Tang, C.-J.; He, J.-L.; Jaing, C.-C.; Liang, C.-J.; Chou, C.-H.; Han, C.-Y.; Tien, C.-L. An All-Solid-State Electrochromic Device Based on $WO_3$–$Nb_2O_5$ Composite Films Prepared by Fast-Alternating Bipolar-Pulsed Reactive Magnetron Sputtering. *Coatings* **2019**, *9*, 9. [CrossRef]

14. Valerini, D.; Hernández, S.; Di Benedetto, F.; Russo, N.; Saracco, G.; Rizzo, A. Sputtered $WO_3$ films for water splitting applications. *Mater. Sci. Semicon. Proc.* **2016**, *42*, 150–154. [CrossRef]

15. Gerand, B.; Nowogrocki, G.; Guenot, J.; Figlarz, M. Structural study of a new hexagonal form of tungsten trioxide. *J. Solid State Chem.* **1979**, *29*, 429–434. [CrossRef]

16. Sun, W.; Yeung, M.T.; Lech, A.T.; Lin, C.-W.; Lee, C.; Li, T.; Duan, X.; Zhou, J.; Kaner, R.B. High surface area tunnels in hexagonal $WO_3$. *Nano Lett.* **2015**, *15*, 4834–4838. [CrossRef]

17. Lee, T.; Lee, Y.; Jang, W.; Soon, A. Understanding the advantage of hexagonal $WO_3$ as an efficient photoanode for solar water splitting: A first-principles perspective. *J. Mater. Chem. A* **2016**, *4*, 11498–11506. [CrossRef]

18. Kollender, J.P.; Gallistl, B.; Mardare, A.I.; Hassel, A.W. Photoelectrochemical water splitting in a tungsten oxide—nickel oxide thin film material library. *Electrochim. Acta* **2014**, *140*, 275–281. [CrossRef]

19. Wiseman, P.J.; Dickens, P.G. The crystal structure of cubic hydrogen tungsten bronze. *J. Solid State Chem.* **1973**, *6*, 374–377. [CrossRef]

20. Mardare, C.C.; Hassel, A.W. Review on the Versatility of Tungsten Oxide Coatings. *Phys. Status Solidi A* **2019**, *216*, 1900047. [CrossRef]

21. Biswas, S.K.; Baeg, J.O. A facile one-step synthesis of single crystalline hierarchical $WO_3$ with enhanced activity for photoelectrochemical solar water oxidation. *Int. J. Hydrog. Energy* **2013**, *8*, 3177–3188. [CrossRef]

22. Gullapalli, S.K.; Vemuri, R.S.; Ramana, C.V. Structural transformation induced changes in the optical properties of nanocrystalline tungsten oxide thin films. *Appl. Phys. Lett.* **2010**, *96*, 171903. [CrossRef]

23. Inoue, T.; Ohtsuka, K.; Yoshida, Y.; Matsuura, Y.; Kajiyama, Y. Metal oxide semiconductor $NO_2$ sensor. *Sens. Actuators B* **1995**, *25*, 388–391. [CrossRef]

24. Kim, T.; Kim, Y.; Yoo, K.S.; Sung, G.; Jung, H. Sensing characteristics of dc reactive sputtered $WO_3$ thin films as an $NO_x$ gas sensor. *Sens. Actuators B Chem.* **2000**, *62*, 102–108. [CrossRef]

25. Qin, Y.X.; Ye, Z.H. DFT study on interaction of $NO_2$ with the vacancy-defected $WO_3$ nanowires for gas-sensing. *Sens. Actuators B Chem.* **2016**, *222*, 499–507. [CrossRef]

26. Di Bartolomeo, E.; Grilli, M.L.; Yoon, J.W.; Traversa, E. Zirconia-Based Electrochemical $NO_x$ Sensors with Semiconducting Oxide Electrodes. *J. Am. Ceram. Soc.* **2004**, *87*, 1883–1889. [CrossRef]

27. Di Bartolomeo, E.; Grilli, M.L.; Traversa, E. Sensing Mechanism of Potentiometric Gas Sensors Based on Stabilized Zirconia with Oxide Electrodes, Is It Always Mixed Potential? *J. Electrochem. Soc.* **2004**, *151*, H133–H139. [CrossRef]

28. Maekawa, T.; Tamaki, J.; Miura, N.; Yamazoe, N. Gold-loaded tungsten-oxide sensor for detection of ammonia in air. *Chem. Lett.* **1992**, *4*, 639–642. [CrossRef]

29. Meixner, H.; Gerblinger, J.; Lampe, U.; Fleischer, M. Thin-film gas sensors based on semiconducting metal oxides. *Sens. Actuators B* **1995**, *23*, 119–125. [CrossRef]

30. Ruokamo, I.; Karkkainen, T.; Huusko, J.; Ruokanen, T.; Blomberg, M.; Torvela, H.; Lantto, V. $H_2S$ response of $WO_3$ thin film sensors manufactured by Silican processing technology. *Sens. Actuators B* **1994**, *19*, 486–488. [CrossRef]

31. Zhu, L.F.; She, J.C.; Luo, J.Y.; Deng, S.Z.; Chen, J.; Xu, N.S. Study of Physical and Chemical Processes of $H_2$ Sensing of Pt-Coated $WO_3$ Nanowire Films. *Phys. Chem. C* **2010**, *114*, 15504–15509. [CrossRef]

32. Shimizu, Y.; Matsunaga, N.; Hyodo, T.; Egashira, M. Improvement of $SO_2$ Sensing Properties of $WO_3$ by Noble Metal Loading. *Sens. Actuators B Chem.* **2011**, *77*, 35–40. [CrossRef]

33. Grilli, M.L.; Di Bartolomeo, E.; Lunardi, A.; Chevallier, L.; Cordiner, S.; Traversa, E. Planar non-nernstian electrochemical sensors: Field test in the exhaust of a spark ignition engine. *Sens. Actuators B Chem.* **2005**, *108*, 319–332. [CrossRef]

34. Grilli, M.L.; Kaabbuathong, N.; Dutta, A.; Di Bartolomeo, E.; Traversa, E. Electrochemical $NO_2$ sensors with $WO_3$ electrodes for high temperature applications. *J. Ceramic Soc. Jpn.* **2002**, *110*, 159–162. [CrossRef]

35. Yamazoe, N.; Shimanoe, K. Theory of Power Laws for Semiconductor Gas Sensors. *Sens. Actuators B Chem.* **2008**, *128*, 566–573. [CrossRef]

36. Wang, C.; Yin, L.; Zhang, L.; Xiang, D.; Gao, R. Metal Oxide Gas Sensors: Sensitivity and Influencing Factors. *Sensors* **2010**, *10*, 2088–2106. [CrossRef]

37. Xu, C.; Tamaki, J.; Miura, N.; Yamazoe, N. Grain size effects on gas sensitivity of porous $SnO_2$-based elements. *Sens. Actuators B* **1991**, *3*, 147–155. [CrossRef]

38. Lin, T.; Lv, X.; Li, S.; Wang, Q. The Morphologies of the Semiconductor Oxides and Their Gas-Sensing Properties. *Sensors* **2017**, *17*, 2779. [CrossRef]

39. Li, X.L.; Lou, T.J.; Sun, X.M.; Li, Y.D. Highly sensitive $WO_3$ hollow-sphere gas sensors. *Inorg. Chem.* **2004**, *43*, 5442–5449. [CrossRef]

40. Yao, Y.; Ji, F.; Yin, M.; Ren, X.; Ma, Q.; Yan, J.; Liu, S.F. Ag Nanoparticle-sensitized $WO_3$ hollow nanosphere for localized surface plasmon enhanced gas sensors. *ACS Appl. Mat. Interface.* **2016**, *8*, 18165–18172. [CrossRef]

41. Kong, W.; Zhang, R.; Zhang, X.; Ji, L.; Yu, G.; Wang, T.; Luo, Y.; Shi, X.; Xu, Y.; Sun, X. $WO_3$ nanosheets rich in oxygen vacancies for enhanced electrocatalytic $N_2$ reduction to $NH_3$. *Nanoscale* **2019**, *11*, 19274–19277. [CrossRef] [PubMed]

42. Liu, Z.; Miyauchi, M.; Yamazaki, T.; Shen, Y. Facile synthesis and $NO_2$ gas sensing of tungsten oxide nanorods assembled microspheres. *Sens. Actuators B* **2009**, *140*, 514–519. [CrossRef]

43. Cao, B.; Chen, J.; Tang, X.; Zhou, W. Growth of monoclinic $WO_3$ nanowire array for highly sensitive $NO_2$ detection. *J. Mater. Chem.* **2009**, *19*, 2323–2327. [CrossRef]

44. Cai, Z.-X.; Li, H.-Y.; Yang, X.-N.; Guo, X. NO sensing by single crystalline $WO_3$ nanowires. *Sens. Actuators B* **2015**, *219*, 346–353. [CrossRef]

45. Ji, H.; Zeng, W.; Li, Y. Gas sensing mechanisms of metal oxide semiconductors: A focus review. *Nanoscale* **2019**, *11*, 22664–22684. [CrossRef]

46. Yan, H.; Zhang, X.; Zhou, S.; Xie, X.; Luo, Y.; Yu, Y. Synthesis of $WO_3$ nanoparticles for photocatalytic $O_2$ evolution by thermal decomposition of ammonium tungstate loading on g-C3N4. *J. Alloys Compounds* **2011**, *509*, L232–L235. [CrossRef]

47. Gomez, C.; Sánchez Martínez, D.; Juarez, I.; Martinez, A.; Torres-Martínez, L. Facile synthesis of m-$WO_3$ powders via precipitation in ethanol solution and evaluation of their photocatalytic activities. *J. Photochem. Photobiol. A: Chem.* **2013**, *262*, 28–33. [CrossRef]

48. Kondalkar, V.V.; Kharade, R.R.; Mali, S.S.; Mane, R.M.; Patil, P.B.; Patil, P.S.; Choudhury, S.; Bhosale, P.N. Nanobrick-like $WO_3$ thin films: Hydrothermalsynthesis and electrochromic application. *Superlattices Microstruct.* **2014**, *73*, 290–295. [CrossRef]

49. Kida, T.; Nishiyama, A.; Hua, Z.; Suematsu, K.; Yuasa, M.; Shimanoe, K. $WO_3$ nanolamella gas sensor: Porosity control using $SnO_2$ nanoparticles for enhanced $NO_2$ sensing. *Langmuir* **2014**, *30*, 2571–2579. [CrossRef]

50. Meng, Z.; Fujii, A.; Hashishin, T.; Wada, N.; Sanada, T.; Tamaki, J.; Kojima, K.; Haneoka, h.; Suzuki, T. Morphological and crystal structural control of tungsten trioxide for highly sensitive $NO_2$ gas sensors. *J. Mater. Chem. C* **2015**, *3*, 1134–1141. [CrossRef]

51. Meulenkamp, E.A. Mechanism of $WO_3$ Electrodeposition from Peroxy-Tungstate Solution. *J. Electrochem. Soc.* **1997**, *144*, 1664–1671. [CrossRef]

52. Breedon, M.; Spizzirri, P.; Taylor, M.; du Plessis, J.; McCulloch, D.; Zhu, J.; Yu, L.; Hu, Z.; Rix, C.; Wlodarski, W.; et al. Synthesis of Nanostructured Tungsten Oxide Thin Films: A Simple, Controllable, Inexpensive, Aqueous Sol−Gel Method. *Cryst. Growth Des.* **2010**, *10*, 430–439. [CrossRef]

53. Yamada, Y.; Tabata, K.; Yashima, T. The character of $WO_3$ film prepared with RF sputtering. *Solar Ener. Mater. Solar Cells* **2007**, *91*, 29–37. [CrossRef]

54. Boyadjiev, S.I.; Georgieva, V.; Stefan, N.; Stan, G.E.; Mihailescu, N.; Visan, A.; Mihailescu, I.N.; Besleaga, C.; Szilágyi, I.M. Characterization of PLD grown $WO_3$ thin films for gas sensing. *Appl. Surf. Sci.* **2017**, *417*, 218–223. [CrossRef]

55. Li, S.; Yao, Z.; Zhou, J.; Zhang, R.; Shen, H. Fabrication and characterization of $WO_3$ thin films on silicon surface by thermal evaporation. *Mater. Lett.* **2017**, *195*, 213–216. [CrossRef]

56. Wang, C.-M.; Wen, C.-Y.; Chen, Y.-C.; Kao, K.-S.; Cheng, D.-L.; Peng, C.-H. Effect of Deposition Temperature on the Electrochromic Properties of Electron Beam-Evaporated $WO_3$ Thin Films. *Integr. Ferroelectr.* **2014**, *158*, 62–68. [CrossRef]

57. Blackman, I.C.S.; Parkin, P. Atmospheric Pressure Chemical Vapor Deposition of Crystalline Monoclinic $WO_3$ and $WO_{3-x}$ Thin Films from Reaction of $WCl_6$ with O-Containing Solvents and Their Photochromic and Electrochromic Properties. *Chem. Mater.* **2005**, *17*, 1583–1590. [CrossRef]

58. Rogers, R.; Seddon, K.; Volkov, S. *Green Industrial Applications of Ionic Liquids*; Springer: Berlin/Heidelberg, Germany, 2002; Volume 818.

59. Holbrey, J.D.; Rogers, R.D. *Ionic Liquids in Synthesis*; Wassercheid, P., Welton, T., Eds.; Wiley-VCH: Weinheim, Germany, 2008; Volume 1, pp. 57–174.

60. Wang, B.; Qin, L.; Mu, T.; Xue, Z.; Gao, G. Are ionic liquids chemically stable? *Chem. Rev.* **2017**, *117*, 7113–7131. [CrossRef]

61. Mahmood, H.; Moniruzzaman, M. Recent Advances of Using Ionic Liquids for Biopolymer Extraction and Processing. *Biotechnol. J.* **2019**, *14*, 1900072. [CrossRef]

62. Verma, C.; Mishra, A.; Chauhan, S.; Verma, P.; Srivastava, V.; Quraishi, M.A.; Ebenso, E.E. Dissolution of cellulose in ionic liquids and their mixed cosolvents: A review. *Sus. Chem. Pharm.* **2019**, *13*, 100162. [CrossRef]

63. Fang, Y.; Ma, P.; Cheng, H.; Tan, G.; Wu, J.; Zheng, J.; Zhou, X.; Fang, S.; Dai, Y.; Lin, Y. Synthesis of Low-Viscosity Ionic Liquids for Application in Dye-Sensitized Solar Cells. *Chem. Asian J.* **2019**, *14*, 4201–4206. [CrossRef] [PubMed]

64. Lobregas, M.O.S.; Camacho, D.H. Gel polymer electrolyte system based on starch grafted with ionic liquid: Synthesis, characterization and its application in dye-sensitized solar cell. *Electrochim. Acta* **2019**, *298*, 219–228. [CrossRef]

65. Dai, C.; Zhang, J.; Huang, C.; Lei, Z. Ionic Liquids in Selective Oxidation: Catalysts and Solvents. *Chem. Rev.* **2017**, *117*, 6929–6983. [CrossRef] [PubMed]

66. Karimi, B.; Tavakolian, M.; Akbari, M.; Mansouri, F. Ionic Liquids in Asymmetric Synthesis: An Overall View from Reaction Media to Supported Ionic Liquid Catalysis. *ChemCatChem* **2018**, *10*, 3173–3205. [CrossRef]

67. Kaur, G.; Sharma, A.; Banerjee, B. Ultrasound and Ionic Liquid: An Ideal Combination for Organic Transformations. *ChemistrySelect* **2018**, *3*, 5283–5295. [CrossRef]

68. Nasrollahzadeh, M.; Motahharifar, N.; Sajjadi, M.; Aghbolagh, A.M.; Shokouhimehr, M.; Varma, R.S. Recent advances in N-formylation of amines and nitroarenes using efficient (nano)catalysts in eco-friendly media. *Green Chem.* **2019**, *21*, 5144–5167. [CrossRef]

69. Sotgiu, G.; Chiarotto, I.; Feroci, M.; Orsini, M.; Rossi, L.; Inesi, A. An electrochemical alternative strategy to the synthesis of β-lactams. Part 3. Room-temperature ionic liquids vs molecular organic solvents. *Electrochim. Acta* **2008**, *53*, 7852–7858. [CrossRef]

70. Feroci, M.; Chiarotto, I.; Inesi, A. Electrolysis of ionic liquids. A possible keystone for the achievement of green solvent-catalyst systems. *Curr. Org. Chem.* **2013**, *17*, 204–219. [CrossRef]

71. D'Anna, F.; Marullo, S.; Vitale, P.; Noto, R. The Effect of the Cation π-Surface Area on the 3D Organization and Catalytic Ability of Imidazolium-Based Ionic Liquids. *Eur. J. Org. Chem.* **2011**, *2011*, 5681–5689. [CrossRef]

72. D'Anna, F.; Marullo, S.; Noto, R. Aryl Azides Formation Under Mild Conditions: A Kinetic Study in Some Ionic Liquid Solutions. *J. Org. Chem.* **2010**, *75*, 767–771. [CrossRef]

73. D'Anna, F.; Marullo, S.; Vitale, P.; Noto, R. Synthesis of aryl azides: A probe reaction to study the synergetic action of ultrasounds and ionic liquids. *Ultrason. Sonochem.* **2012**, *19*, 136–142. [CrossRef] [PubMed]

74. Marullo, S.; D'Anna, F.; Rizzo, C.; Noto, R. The ultrasounds–ionic liquids synergy on the copper catalyzed azide–alkyne cycloaddition between phenylacetylene and 4-azidoquinoline. *Ultrason. Sonochem.* **2015**, *23*, 317–323. [CrossRef] [PubMed]

75. Rizzo, C.; D'Anna, F.; Marullo, S.; Noto, R. Task Specific Dicationic Ionic Liquids: Recyclable Reaction Media for the Mononuclear Rearrangement of Heterocycles. *J. Org. Chem.* **2014**, *79*, 8678–8683. [CrossRef] [PubMed]

76. Pandolfi, F.; Feroci, M.; Chiarotto, I. Role of anion and cation in the 1-methyl-3-butylimidazolium ionic liquids BMImX: The Knoevenagel condensation. *ChemistrySelect* **2018**, *3*, 4745–4749. [CrossRef]

77. Pandolfi, F.; Chiarotto, I.; Mattiello, L.; Petrucci, R.; Feroci, M. Two different selective ways in the deprotonation of β-bromopropionanilides: β-lactams or acrylanilides formation. *ChemistrySelect* **2019**, *4*, 12871–12874. [CrossRef]

78. Chiarotto, I.; Feeney, M.M.M.; Feroci, M.; Inesi, A. Electrogenerated N-heterocyclic carbene: N-acylation of chiral oxazolidin-2-ones in ionic liquids. *Electrochim. Acta* **2009**, *54*, 1638–1644. [CrossRef]

79. Chiarotto, I.; Feroci, M.; Orsini, M.; Sotgiu, G.; Inesi, A. Electrogenerated N-heterocyclic carbene: N-functionalization of benzoxazolones. *Tetrahedron* **2009**, *65*, 3704–3710. [CrossRef]

80. Feroci, M.; Elinson, M.N.; Rossi, L.; Inesi, A. The double role of ionic liquids in organic electrosynthesis: Precursors of N-heterocyclic carbenes and green solvents. Henry reaction. *Electrochem. Commun.* **2009**, *11*, 1523–1526. [CrossRef]

81. Feroci, M.; Chiarotto, I.; Vecchio Ciprioti, S.; Inesi, A. On the reactivity and stability of electrogenerated N-heterocyclic carbene in parent 1-butyl-3-methylimidazolium tetrafluoroborate: Formation and use of N-heterocyclic carbene-$CO_2$ adduct as latent catalyst. *Electrochim. Acta* **2013**, *109*, 95–101. [CrossRef]

82. Feroci, M.; Chiarotto, I.; D'Anna, F.; Gala, F.; Noto, R.; Ornano, L.; Zollo, G.; Inesi, A. N-Heterocyclic carbenes and parent cations: Acidity, nucleophilicity, stability, and hydrogen bonding-electrochemical study and ab initio calculations. *ChemElectroChem* **2016**, *3*, 1133–1141. [CrossRef]

83. Chiarotto, I.; Mattiello, L.; Pandolfi, F.; Rocco, D.; Feroci, M. NHC in imidazolium acetate ionic liquids: Actual or potential presence? *Front. Chem.* **2018**, *6*, 355. [CrossRef] [PubMed]

84. Ali, F.I.M.; Awwad, F.; Greish, Y.E.; Abu-Hani, A.F.S.; Mahmoud, S.T. Fabrication of low temperature and fast response $H_2S$ gas sensor based on organic-metal oxide hybrid nanocomposite membrane. *Org. Electron.* **2020**, *76*, 105486. [CrossRef]

85. Bahadori, M.; Tangestaninejad, S.; Bertmer, M.; Moghadam, M.; Mirkhani, V.; Mohammadpoor−Baltork, I.; Kardanpour, R.; Zadehahmadi, F. Task-Specific Ionic Liquid Functionalized−MIL−101(Cr) as a Heterogeneous and Efficient Catalyst for the Cycloaddition of $CO_2$ with Epoxides Under Solvent Free Conditions. *ACS Sustain. Chem. Eng.* **2019**, *7*, 3962–3973. [CrossRef]

86. Eyckens, D.J.; Champion, M.E.; Fox, B.L.; Yoganantharajah, P.; Gibert, Y.; Welton, T.; Henderson, L.C. Solvate Ionic Liquids as Reaction Media for Electrocyclic Transformations. *Eur. J. Org. Chem.* **2016**, *2016*, 913–917. [CrossRef]

87. Eyckens, D.J.; Demir, B.; Walsh, T.R.; Welton, T.; Henderson, L.C. Determination of Kamlet–Taft parameters for selected solvate ionic liquids. *Phys. Chem. Chem. Phys.* **2016**, *18*, 13153–13157. [CrossRef]

88. Lui, M.Y.; Crowhurst, L.; Hallett, J.P.; Hunt, P.A.; Niedermeyer, H.; Welton, T. Salts dissolved in salts: Ionic liquid mixtures. *Chem. Sci.* **2011**, *2*, 1491–1496. [CrossRef]

89. Dupont, J. From Molten Salts to Ionic Liquids: A "Nano" Journey. *Acc. Chem. Res.* **2011**, *44*, 1223–1231. [CrossRef]

90. Dupont, J.; Suarez, P.A.Z. Physico-chemical processes in imidazolium ionic liquids. *Phys. Chem. Chem. Phys.* **2006**, *8*, 2441–2452. [CrossRef]

91. Carvalho, T.O.; Carvalho, P.H.P.R.; Correa, J.R.; Guido, B.C.; Medeiros, G.A.; Eberlin, M.N.; Coelho, S.E.; Domingos, J.B.; Neto, B.A.D. Palladium Catalyst with Task-Specific Ionic Liquid Ligands: Intracellular Reactions and Mitochondrial Imaging with Benzothiadiazole Derivatives. *J. Org. Chem.* **2019**, *84*, 5118–5128. [CrossRef]

92. Qian, W.; Tan, X.; Su, Q.; Cheng, W.; Xu, F.; Dong, L.; Zhang, S. Transesterification of Isosorbide with Dimethyl Carbonate Catalyzed by Task-Specific Ionic Liquids. *ChemSusChem* **2019**, *12*, 1169–1178. [CrossRef]

93. Xie, W.; Shao, L. Phosphorylation of Corn Starch in an Ionic Liquid. *Starch/Stärke* **2009**, *61*, 702–708. [CrossRef]

94. El Sayed, S.; Bordet, A.; Weidenthaler, C.; Hetaba, W.; Luska, K.L.; Leitner, W. Selective Hydrogenation of Benzofurans Using Ruthenium Nanoparticles in Lewis Acid-Modified Ruthenium-Supported Ionic Liquid Phases. *ACS Catal.* **2020**, *10*, 2124–2130. [CrossRef]

95. Fatehi, A.; Ghorbani-Vaghei, R.; Alavinia, S.; Mahmoodi, J. Synthesis of Quinazoline Derivatives Catalyzed by a New Efficient Reusable Nanomagnetic Catalyst Supported with Functionalized Piperidinium Benzene-1,3-Disulfonate Ionic Liquid. *ChemistrySelect* **2020**, *5*, 944–951. [CrossRef]

96. Zhao, Q.; Yang, C.; Fang, M.; Jiang, T. Performance of Brönsted-Lewis acidic ionic liquids supported Ti-SBA-15 for the esterification of acetic acid to benzyl alcohol. *Appl. Catal. A General* **2020**, *594*, 117470. [CrossRef]

97. Billeci, F.; D'Anna, F.; Gunaratne, H.Q.N.; Plechkova, N.V.; Seddon, K.R. "Sweet" ionic liquid gels: Materials for sweetening of fuels. *Green Chem.* **2018**, *20*, 4260–4276. [CrossRef]

98. Guo, P.; Su, A.; Wei, Y.; Liu, X.; Li, Y.; Guo, F.; Li, J.; Hu, Z.; Sun, J. Healable, Highly Conductive, Flexible, and Nonflammable Supramolecular Ionogel Electrolytes for Lithium-Ion Batteries. *ACS Appl. Mater. Interfaces* **2019**, *11*, 19413–19420. [CrossRef] [PubMed]

99. Kuddushi, M.; Mata, J.; Malek, N. Self-Sustainable, self-healable, Load Bearable and Moldable stimuli responsive ionogel for the Selective Removal of Anionic Dyes from aqueous medium. *J. Mol. Liq.* **2020**, *298*, 112048. [CrossRef]

100. Marullo, S.; Rizzo, C.; Dintcheva, N.T.; Giannici, F.; D'Anna, F. Ionic liquids gels: Soft materials for environmental remediation. *J. Colloid Interface Sci.* **2018**, *517*, 182–193. [CrossRef]

101. Rizzo, C.; Marullo, S.; Campodonico, P.R.; Pibiri, I.; Dintcheva, N.T.; Noto, R.; Millan, D.; D'Anna, F. Self-Sustaining Supramolecular Ionic Liquid Gels for Dye Adsorption. *ACS Sustain. Chem. Eng.* **2018**, *6*, 12453–12462. [CrossRef]

102. Billeci, F.; D'Anna, F.; Feroci, M.; Cancemi, P.; Feo, S.; Forlino, A.; Tonnelli, F.; Seddon, K.R.; Gunaratne, H.Q.N.; Plechkova, N.V. When Functionalization Becomes Useful: Ionic Liquids with a "Sweet" Appended Moiety Demonstrate Drastically Reduced Toxicological Effects. *ACS Sustain. Chem. Eng.* **2020**, *8*, 926–938. [CrossRef]

103. Egorova, K.S.; Ananikov, V.P. Toxicity of Ionic Liquids: Eco(cyto)activity as Complicated, but Unavoidable Parameter for Task-Specific Optimization. *ChemSusChem* **2014**, *7*, 336–360. [CrossRef] [PubMed]

104. Hulsbosch, J.; De Vos, D.E.; Binnemans, K.; Ameloot, R. Biobased Ionic Liquids: Solvents for a Green Processing Industry? *ACS Sustain. Chem. Eng.* **2016**, *4*, 2917–2931. [CrossRef]

105. Socha, A.M.; Parthasarathi, R.; Shi, J.; Pattathil, S.; Whyte, D.; Bergeron, M.; George, A.; Tran, K.; Stavila, V.; Venkatachalam, S.; et al. Efficient biomass pretreatment using ionic liquids derived from lignin and hemicellulose. *Proc. Natl. Acad. Sci. USA* **2014**, *111*, E3587–E3595. [CrossRef] [PubMed]

106. Worlds Health Organization (WHO). Health Topics. Available online: https://www.who.int/health-topics/air-pollution#tab=tab_1 (accessed on 14 February 2020).

107. Worlds Health Organization (WHO). Available online: https://www.who.int/airpollution/ambient/pollutants/en/) (accessed on 14 February 2020).

108. Duk-Dong, L.; Dae-Sik, L. Environmental gas sensors. *IEEE Sensors J.* **2001**, *1*, 214–224. [CrossRef]

109. Rai, A.C.; Kumar, P.; Pilla, F.; Skouloudis, A.N.; Di Sabatino, S.; Ratti, C.; Yasar, A.; Rickerby, D. End-user perspective of low-cost sensors for outdoor air pollution monitoring. *Sci. Total Environ.* **2017**, *607–608*, 691–705. [CrossRef]

110. Pang, X.; Shaw, M.D.; Gillot, S.; Lewis, A.C. The impacts of water vapour and co-pollutants on the performance of electrochemical gas sensors used for air quality monitoring. *Sens. Actuators B Chem.* **2018**, *266*, 674–684. [CrossRef]

111. Moseley, P.T. Progress in the development of semiconducting metal oxide gas sensors: A review. *Meas. Sci. Technol.* **2017**, *28*, 082001. [CrossRef]

112. Dong, C.; Zhao, R.; Yao, L.; Ran, Y.; Zhang, X.; Wang, Y. A review on $WO_3$ based gas sensors: Morphology control and enhanced sensing properties. *J. Alloys Compd.* **2020**, *820*, 153194. [CrossRef]

113. Hariharan, V.; Gnanavel, B.; Sathiyapriya, R.; Aroulmoji, V.A. A Review on Tungsten Oxide ($WO_3$) and their Derivatives for Sensor Applications. *Int. J. Adv. Sci. Eng.* **2019**, *5*, 1163–1168. [CrossRef]

114. Sari, W.P.; Leigh, S.; Covington, J. Tungsten Oxide Based Sensor for Oxygen Detection. *Proceedings* **2018**, *2*, 952. [CrossRef]

115. Staerz, A.; Somacescu, S.; Epifani, M.; Russ, T.; Weimar, U.; Barsan, N. $WO_3$ Based Gas Sensors. *Proceedings* **2018**, *2*, 826. [CrossRef]

116. Li, Z.; Li, J.; Song, L.; Gong, H.; Niu, Q. Ionic liquid-assisted synthesis of $WO_3$ particles with enhanced gas sensing properties. *J. Mat. Chem. A* **2013**, *1*, 15377–15382. [CrossRef]

117. Zhang, Y.; Cheng, X.; Zhang, X.; Major, Z.; Xu, Y.; Gao, S.; Zhao, H.; Huo, L. Ionic liquid-assisted synthesis of tungsten oxide nanoparticles with enhanced $NO_2$ sensing properties at near room temperature. *Appl. Surf. Sci.* **2020**, *505*, 144533. [CrossRef]

118. Favard, A.Y.X.; Anguille, S.; Moulin, P.; Seguin, J.-L.; Aguir, K.; Bendahan, M. Ionic Liquids Filter for Humidity Effect Reduction on Metal Oxide Gas Sensor Response. *Sensors Transduc.* **2018**, *222*, 6–11.

119. Abu-Hani, A.F.S.; Awwad, F.; Greish, Y.E.; Ayesh, A.I.; Mahmoud, S.T. Design, fabrication, and characterization of low-power gas sensors based on organic-inorganic nano-composite. *Org. Electron.* **2017**, *42*, 284–292. [CrossRef]

120. Daves, W.; Ersoez, B. Electrochemical Gas Sensor. International Patent Application No. WO2018/234185 A1, 27 December 2018.

121. Bhutto, A.W.; Abro, R.; Gao, S.; Abbas, T.; Chen, X.; Yu, G. Oxidative desulfurization of fuel oils using ionic liquids: A review. *J. Taiwan Inst. Chem. Eng.* **2016**, *62*, 84–97. [CrossRef]

122. Daage, M.; Chianelli, R.R. Structure-Function Relations in Molybdenum Sulfide Catalysts: The "Rim-Edge" Model. *J. Catal.* **1994**, *149*, 414–427. [CrossRef]
123. Topsøe, H. The role of Co–Mo–S type structures in hydrotreating catalysts. *Appl. Catal. A General* **2007**, *322*, 3–8. [CrossRef]
124. Shan, S.; Yuan, P.; Han, W.; Shi, G.; Bao, X. Supported NiW catalysts with tunable size and morphology of active phases for highly selective hydrodesulfurization of fluid catalytic cracking naphtha. *J. Catal.* **2015**, *330*, 288–301. [CrossRef]
125. Hossain, M.N.; Park, H.C.; Choi, H.S. A Comprehensive Review on Catalytic Oxidative Desulfurization of Liquid Fuel Oil. *Catalyst* **2019**, *9*, 12. [CrossRef]
126. Sun, H.; Wu, P.; He, J.; Liu, M.; Zhu, L.; Zhu, F.; Chen, G.; He, M.; Zhu, W. Fabrication of oxygen-defective tungsten oxide nanorods for deep oxidative desulfurization of fuel. *Pet. Sci.* **2018**, *15*, 849–856. [CrossRef]
127. Rongxiang, Z.; Xiuping, L.; Jianxun, S.; Weiwei, S.; Xiaohan, G. Preparation of WO$_3$/C Composite and Its Application in Oxidative Desulfurization of Fuel. *China Pet. Process. Petrochem. Technol.* **2017**, *19*, 65–73.
128. Li, X.; Zhao, R.; Mao, C. Polycrystalline Phase WO$_3$/g-C$_3$N$_4$ as a High Efficient Catalyst for Removal of DBT in Model Oil. *China Pet. Process. Petrochem. Technol.* **2019**, *21*, 36–45.
129. Zhang, M.; Zhu, W.; Li, H.; Xun, S.; Li, M.; Li, Y.; Wei, Y.; Li, H. Fabrication and characterization of tungsten-containing mesoporous silica for heterogeneous oxidative desulfurization. *Chin. J. Catal.* **2016**, *37*, 971–978. [CrossRef]
130. Xun, S.; Hou, C.; Li, H.; He, M.; Ma, R.; Zhang, M.; Zhu, W.; Li, H. Synthesis of WO$_3$/mesoporous ZrO$_2$ catalyst as a high-efficiency catalyst for catalytic oxidation of dibenzothiophene in diesel. *J. Mat. Sci.* **2018**, *53*, 15927–15938. [CrossRef]
131. Ma, R.; Guo, J.; Wang, D.; He, M.; Xun, S.; Gu, J.; Zhu, W.; Li, H. Preparation of highly dispersed WO$_3$/few layer g-C$_3$N$_4$ and its enhancement of catalytic oxidative desulfurization activity. *Colloid. Surface. A: Physicochem. Eng. Aspects* **2019**, *572*, 250–258. [CrossRef]
132. Wang, C.; Li, A.; Xu, J.; Wen, J.; Zhang, H.; Zhang, L. Preparation of WO$_3$/CNT catalysts in presence of ionic liquid [C$_{16}$mim]Cl and catalytic efficiency in oxidative desulfurization. *J. Chem. Technol. Biotechnol.* **2019**, *94*, 3403–3412. [CrossRef]

*Review*

# Design, Fabrication, and Characterization of New Materials Based on Zirconia Doped with Mixed Rare Earth Oxides: Review and First Experimental Results

Adrian Mihail Motoc [1,*], Sorina Valsan [1,*], Anca Elena Slobozeanu [1,*], Mircea Corban [1], Daniele Valerini [2,*], Mythili Prakasam [3], Mihail Botan [4], Valentin Dragut [1], Bogdan St. Vasile [5], Adrian Vasile Surdu [5], Roxana Trusca [5], Maria Luisa Grilli [6,*] and Robert Radu Piticescu [1,*]

[1] National R&D Institute for Nonferrous and Rare Metals-IMNR, Laboratory for Advanced and Nanostructured Materials and HighPTMET Center, 102 Biruintei Blvd, Pantelimon, 077145 Ilfov, Romania; corban.mircea@imnr.ro (M.C.); dragutv@imnr.ro (V.D.)

[2] ENEA–Italian National Agency for New Technologies, Energy and Sustainable Economic Development, Brindisi Research Centre, S.S.7 Appia-km 706, 72100 Brindisi, Italy

[3] University Bordeaux, CNRS, Bordeaux INP, ICMCB, UMR 5026 Pessac, 33600 Bordeaux, France; mythili.prakasam@ubordeaux.fr

[4] National R&D Institute for Aeronautics "Elie Carafoli", 066126 Bucharest, Romania; botan.mihail@incas.ro

[5] National Research Center for Micro and Nanomaterials, University Politehnica from Bucharest, 060042 Bucharest, Romania; bogdan.vasile@upb.ro (B.S.V.); adrian.surdu@upb.ro (A.V.S.); roxanatrusca@yahoo.com (R.T.)

[6] ENEA-Italian National Agency for New Technologies, Energy and Sustainable Economic Development, Casaccia Research Centre, Via Anguillarese 301, 00123 Rome, Italy

[*] Correspondence: amotoc@imnr.ro (A.M.M.); svalsan@imnr.ro (S.V.); a.slobozeanu@imnr.ro (A.E.S.); daniele.valerini@enea.it (D.V.); marialuisa.grilli@enea.it (M.L.G.); rpiticescu@imnr.ro (R.R.P.); Tel.: +402-135-220-48 (A.M.M.)

Received: 29 April 2020; Accepted: 29 May 2020; Published: 3 June 2020

**Abstract:** Monazite is one of the most valuable natural resources for rare earth oxides (REOs) used as dopants with high added value in ceramic materials for extreme environments applications. The complexity of the separation process in individual REOs, due to their similar electronic configuration and physical–chemical properties, is reflected in products with high price and high environmental footprint. During last years, there was an increasing interest for using different mixtures of REOs as dopants for high temperature ceramics, in particular for $ZrO_2$-based thermal barrier coatings (TBCs) used in aeronautics and energy co-generation. The use of mixed REOs may increase the working temperature of the TBCs due to the formation of tetragonal and cubic solid solutions with higher melting temperatures, avoiding grain size coarsening due to interface segregation, enhancing its ionic conductivity and sinterability. The thermal stability of the coatings may be further improved by using rare earth zirconates with perovskite or pyrochlore structures having no phase transitions before melting. Within this research framework, firstly we present a review analysis about results reported in the literature so far about the use of $ZrO_2$ ceramics doped with mixed REOs for high temperature applications. Then, preliminary results about TBCs fabricated by electron beam evaporation starting from mixed REOs simulating the real composition as occurring in monazite source minerals are reported. This novel recipe for $ZrO_2$-based TBCs, if optimized, may lead to better materials with lower costs and lower environmental impact, as a result of the elimination of REOs extraction and separation in individual lanthanides. Preliminary results on the compositional, microstructure, morphological, and thermal properties of the tested materials are reported.

**Keywords:** zirconia; rare earth zirconates; thermal barrier coatings; microstructure characterization; thermal shock resistance

## 1. Introduction

Necessity for better performance and efficiency in applications across stationary power plants, aerospace and automotive industries has led to the development of thermal barrier coatings (TBCs) and multi-layer coating systems over the last 50 years [1]. During recent years it was demonstrated that co-doping of zirconia ceramics with rare earth oxides (REOs) may avoid grain size coarsening due to interface segregation, enhancing its ionic conductivity and sinterability [2,3]. The co-doping of zirconia with different REOs was reported to improve the thermal properties of thermal barrier coatings and oxidation properties due to reduction of mechanical stresses and porosity in the oxide layers [4,5]. These improvements play a crucial role in the segments mentioned above because they are effective means for protecting the hot parts against the effects of high temperature in corrosive/erosive environments, extending the life of the metallic parts and reducing the maintenance costs [6–9].

Typically, TBCs are made up of different successive layers deposited on a substrate (typically made of a super alloy): the metal bond coat (BC), the thermally-grown oxide (TGO) generated at the interface by oxidation of the bonding layer, and the upper layer (top coat) in direct contact with the environment [10]. In order to operate under extreme conditions, an effective TBC must meet certain requirements, which severely limits the number of materials that can be used. Low thermal conductivity, high melting point, phase stability, good adhesion to the substrate, chemical inertia, thermodynamic compatibility with the metallic substrate and TGO, high thermal cycling resistance are the major requirements to be achieved [6,11].

Ceramic materials are usually capable of satisfying the required properties. Compared to metallic materials, ceramic coatings offer several advantages due to the high hardness and wear resistance when exposed to thermal and corrosive conditions, coupled with relatively low densities [12]. Flame-sprayed $Al_2O_3$/CaO-doped $ZrO_2$ ceramic coatings were firstly developed. However, they did not prove to be viable materials for the more advanced thermal barrier applications due to relatively high thermal conductivity and phase transitions of $Al_2O_3$, leading to shrinkage and the associated cracking effect on coating life [7,13].

Yttria-doped zirconia ($ZrO_2$ doped with 7–8 wt.% $Y_2O_3$) are presently considered as a "gold standard" in coating materials for TBC applications [14]. Compared to other ceramics, zirconia has good corrosion / erosion resistance, lower intrinsic thermal conductivity and a coefficient of thermal expansion best suited [6]. Table 1 illustrates some important properties of the traditional TBC layers.

**Table 1.** Properties of thermal barrier coatings (TBC).

| TBC Properties | | |
|---|---|---|
| $Al_2O_3$ (TGO) | NiCoCrAlY | 8YSZ |
| $T_m$ ~2323 K<br>$\lambda = 5.8$ W m$^{-1}$·K$^{-1}$ (1400 K)<br>$\alpha = 9.6 \times 10^{-6}$ K$^{-1}$ (1273 K)<br>E = 30 GPa (293 K)<br>$\upsilon = 0.26$ | $T_m$ ~1863 K<br>$\lambda = 320$ W m$^{-1}$·K$^{-1}$ (293 K)<br>$\alpha = 10.7 \times 10^{-6}$ K$^{-1}$<br>(293–1273 K)<br>E = 40 GPa (293 K)<br>$\upsilon = 0.22$ | $T_m$ ~2873 K<br>$\lambda$~2.5 W m$^{-1}$·K$^{-1}$ (298 K)<br>$\alpha = 10.7 \times 10^{-6}$ K$^{-1}$<br>(293–1273 K) E = 40 GPa<br>(293 K)<br>$\upsilon = 0.22$ |

Symbols in the table have the following meanings: TGO—thermally grown oxide on bond coat; $T_m$—melting point; $\lambda$—thermal conductivity; $\alpha$—thermal expansion coefficient; E—Young's modulus; $\upsilon$—Poisson's number.

The stable form of pure $ZrO_2$ is the monoclinic phase, present at temperatures below 1100 °C. Between 1100 and 2370 °C $ZrO_2$ crystallizes in the tetragonal phase, and the cubic phase is found between 2370 °C and 2706 °C (the melting point) [14,15]. The mechanical properties of zirconia are drastically affected by microcracking caused by the transition from the tetragonal to the monoclinic phase, which occurs with an increase in volume of about 4%. Dopants such as MgO, CaO, $Y_2O_3$, and other rare earth oxides are compulsory to stabilize both the tetragonal and cubic phases at room temperature [6,15]. These ceramic materials have very broad and interesting range of electrical, mechanical, thermal, optical and biocompatibility properties, which result from the structural and the intrinsic chemical-physical properties listed in Table 1 [15]. $Y_2O_3$-stabilized $ZrO_2$ (YSZ) was found to be the most suitable material for TBC applications, with an optimum amount of $Y_2O_3$ around 7–8 wt.% (4–4.5 mol%). This composition offers a high degree of resistance to spallation and excellent thermal stability as presented in Table 2 [16,17].

**Table 2.** $ZrO_2$ properties [16].

| $ZrO_2$ **Properties** | | | |
|---|---|---|---|
| Melting point, °C | 2700 | Evaporation rate $\mu$m/h $\times 10^{-5}$ at 1650 °C | 670 |
| Density, g/cm$^3$ | 5.6 | at 1927 °C | 0.75 |
| Vapor pressure, Pa $\times 10^{-6}$ at 1650 °C | 10.6 | at 2200 °C | 44 |
| $\times 10^{-5}$ at 1927 °C | 127 | Oxygen permeability g/cm$^{-1}$ s$^{-1}$ $\times 10^{-13}$ at 1000 °C | 120 |
| $\times 10^{-3}$ at 2200 °C | 78.7 | $\times 10^{-11}$ at 1400 °C | 60 |
| Crystal structure below 1170 °C | Monoclinic | $\times 10^{-10}$ at 1800 °C | 30 |
| 1170–2370 °C | Tetragonal | Limit of stability with carbon, °C | 1600 |
| 2370–2706 °C | Cubic | Thermal expansion, ppm/°C | 7.5 |

However, at temperatures above 1200 °C in the long term, YSZ loses its phase stability but the main critical issue is calcium–magnesium–alumino–silicate (CMAS) attack that lowers the thermal insulating power [18,19]. Therefore, it is desired to develop new TBCs that can operate at a temperature above 1200 °C. An actual research topic is doping $ZrO_2$ ceramics with different rare earth oxides due to their high temperature transformation and the low thermal conductivity, which allow to increase the operating temperatures of turbines.

## 2. Synthesis and Properties of $ZrO_2$ Ceramic Powders Doped with Mixed REOs

Different methods were proposed for the synthesis of zirconia doped with mixed rare earth oxides. We report in the following about different synthesis routes, the different $ZrO_2$ dopants and different applications, with focus on TBCs.

$ZrO_2$ thin films doped with 8 wt.% rare earths (Ce, Gd, and Y) were prepared by Sasikumar et al. on glass substrates by sol-gel, followed by spin-coating technique and annealed at 600 °C. X-ray diffraction (XRD) analysis revealed that for non-doped $ZrO_2$ films, only the monoclinic phase is present, while for the doped $ZrO_2$ films both monoclinic and tetragonal phase appear. Because of doping with RE ions, the crystallite size decreased with the increase of oxygen vacancies. Scanning electron microscopy (SEM) images exhibited the rod-shaped and square-shaped grains for the undoped $ZrO_2$ and Gd:$ZrO_2$ thin films, respectively [20].

Ce-doped $ZrO_2$ and Dy-doped $ZrO_2$ were synthesized by Mekala and co-workers by a coprecipitation method. The XRD study revealed that the synthesized nanoparticles showed tetragonal (t-$ZrO_2$) and monoclinic (m-$ZrO_2$) phase and also confirmed the decrease in crystal size with Dy and Ce doping. Apparently, doping with RE had different effects on the size and morphology of $ZrO_2$ nanopowders. $ZrO_2$ doped with Ce had the smallest particle size (6 nm), doping with Dy led to particle size of 9 nm, and the non-doped $ZrO_2$ size was about 12 nm. Authors stated that Ce doped sample has nano-flakes like morphology, Dy doping has snow-flakes like morphology and undoped has cauliflower-like morphology. This result reveals that Dy doping promotes, and Ce doping inhibits the grain growth and crystallization of $ZrO_2$, which was consistent with the result of XRD analysis [21].

To understand the thermal behavior of redox ability for different ceria-zirconia based materials during thermal aging process, Deng and collaborators have sintered five $CeO_2$–$ZrO_2$–$Y_2O_3$–$La_2O_3$ quaternary mixed oxides (CZ) with varied ratio of Ce/Zr. The Ce-rich $CeO_2$–$ZrO_2$ based oxide (Ce/Zr >1) exhibits a declined oxygen storage capacity (OSC) and deteriorated reducibility after thermal aging. Raman and X-Ray Photoelectron Spectroscopy (XPS) analyses suggest that the abundant available Zr ions in Zr-rich samples do not incorporate into ceria lattices completely during moderate calcination, and that these remaining Zr ions migrate into ceria lattices further during thermal aging accompanied by sintering. This is considered as the most reasonable explanation for the enhanced OSC of Zr-rich samples after thermal aging [22].

$ZrO_2$ powders doped with 9.5% $Y_2O_3$, 5.6% $Yb_2O_3$, and 5.2% $Gd_2O_3$ have been synthesized by a chemical co-precipitation method with improved thermal stability at 1300 °C for 50 h. The higher phase's stability was given by the increased concentration of stabilizers leading to the increase of the concentration of oxygen vacancies, thus decreasing the tetragonality [14].

Madhusudhana et al. synthesized Gd (1–9 mol%) doped $ZrO_2$ nanocrystals by solution combustion method using Glycine as a fuel, furnace temperature being 400 °C. XRD analysis showed high crystallinity of the sample, with particle size of 25–35 nm. It was observed that the addition of $Gd^{3+}$ ions influences the change of the structure of $ZrO_2$ grains from flaky structures at 1 mol% of Gd, to uniform spherical structures at 9 mol% of Gd, creating more oxygen vacancies. AC impedance spectroscopy results of the as synthesized Gd doped $ZrO_2$ show good dielectric properties with a very high dielectric constant ($\varepsilon' = 345$) and as well a high AC conductivity ($\sigma_{ac} = 0.06837$ S·cm$^{-1}$) at 10 MHz for 7 mol% of Gd doped $ZrO_2$. This material is best suitable for solid oxide fuel cell applications [23].

Xiao and co-workers systematically investigated the band gap, the interfacial properties and the optoelectronic properties of Gd (0.5, 10, 15 mol%) doped $ZrO_2$ high-k gate dielectric films deposited by solution method, reaching the conclusion that the incorporation of Gd has a positive effect on optimizing $ZrO_2$ dielectrics applied in further Complementary Metal Oxide Semiconductor (CMOS) devices. They found that the band gap of sample increases with the increase of Gd concentration, conduction band offset increases from 2.57 to 3.06 eV with the increase of Gd concentration, indicating that the band offset is larger than the minimum requirement for the barrier height of over 1 eV for future CMOS. The increase of dielectric constant (k) from 8.5 to 10.3 with the Gd concentration increase from 0 to 10% may be attributed to the Gd incorporation, which is favorable for improving the interfacial quality and prevent the growth of low-k SiOx interlayer. When the Gd concentration increases to 15%, the k value decreases, which is probably due to the lower dielectric constant of $Gd_2O_3$ comparing with $ZrO_2$. The negative flat-band voltage values suggest that there are positive oxide charges in the film, which may be attributed to the oxygen vacancies and positive fixed charges in the film and near the interface. The incorporation of Gd also decreases leakage current density from $3.2 \times 10^{-5}$ to $1.8 \times 10^{-6}$ A/cm$^2$ [24].

Lei Guo and co-workers fabricated by co-precipitation and calcination method 1 mol% $RE_2O_3$ (RE = La, Nd, Gd, and Yb) and 1 mol% $Yb_2O_3$ co-doped YSZ (1RE1Yb-YSZ), in order to obtain improved phase stability and reduced thermal conductivity. For 1RE1Yb-YSZ ceramics, the phase stability of the metastable tetragonal phase (t′) increased with decreasing size of $RE^{3+}$, mainly attributed to the reduced driving force for the partitioning of the t′ phase. By codoping, the thermal conductivity was lower than that of YSZ, with the exception of the sample 1La1Yb-YSZ, which showed an undesirable high thermal conductivity, being attributable to the high content of phase m due to the decomposition of phase t′. Authors stated that, considering the full ceramic properties, 1Gd1Yb-YSZ could be a good potential material for TBC applications [25].

In their study, Xiwen Song et al. investigated the influences of the partial substitution of $Y_2O_3$ with the equivalent $Ln_2O_3$ (Ln = Nd, Sm, and Gd) on the phase structure and thermal properties of $ZrO_2$–$Nb_2O_5$–$Y_2O_3$ ceramics. The powders for $ZrO_2$–$Nb_2O_5$–$Y_2O_3$–$Ln_2O_3$ samples were synthesized by solid-state reaction of well-mixed stable oxides. All $ZrO_2$–$Nb_2O_5$–$Y_2O_3$–$Ln_2O_3$ compositions consist of a single non-transformable tetragonal phase with an exceptionally high tetragonality. However, partial replacement of $Y^{3+}$ with $Ln^{3+}$ results in a slight decrease in tetragonality. These compositions have low thermal conductivity. The composition with $Nd^{3+}$ replaced with $Y^{3+}$ showed the lowest thermal conductivity of ~1.69 W m$^{-1}$ K$^{-1}$, approximately 30% lower than that of composition $ZrO_2$–$Nb_2O_5$–$Y_2O_3$ [26].

Xiwen Song and co-workers determined the effect of rare earth oxides $Ln_2O_3$ (Ln = Nd, Sm, and Gd) on the structure and thermal properties of $ZrO_2$-based ceramics doped with 8.3 mol% $Ta_2O_5$ + 8.3 mol% $Y_2O_3$ by solid-state reaction. When 1 mol% $Ln_2O_3$ substituted 1 mol%

$Y_2O_3$, all composites showed only the tetragonal phase. The results of the study suggest that the addition of $Ln_2O_3$ may be a viable strategy for decreasing grain growth, specific heat capacity and thermal conductivity, which is beneficial in improving the performance of the $ZrO_2$–$Ta_2O_5$–$Y_2O_3$ system for TBC applications. The most effective oxide in reducing the thermal conductivity of the system was $Gd_2O_3$ [27].

Mikhailov and al. obtained fine-grained ceramics based on solid solutions of $ZrO_2$–x–$Ln_2O_3$ (where Ln = Sm and Yb). These were obtained by the colloidal chemical synthesis of the powders followed by SPS (spark plasma sintering). They showed that the addition of lanthanides in zirconium efficiently stabilizes the structure and no phase transformation occurs during heating, ensuring a high density of the ceramic obtained by SPS and high hydrolytic stability. They also demonstrated that the introduction of lanthanides leads to a higher intensity of contraction and reduces the activation energy of ceramic sintering based on $ZrO_2$. The influence of lanthanide oxides on the sintering activation energy of ceramics in SPS is determined by the influence of ions of $Sm^{3+}$ and $Yb^{3+}$ on the diffusion properties of grain boundaries and by the proportion between the atom radius of zirconium and lanthanides [28].

Bahamirian developed a new material, ZGYbY powder: $ZrO_29.5Y_2O_35.6Yb_2O_35.2Gd_2O_3$ synthesized by the chemical precipitation method, in order to increase the stability of the YSZ phases at higher temperatures. During the thermal cycle at 1100 °C for 50 h they confirmed the stability of the phases of both powders, ZGYbY and YSZ. The results of the thermal cycle at 1300 °C for 50 h indicated that the YSZ powder decomposed into two new phases, including cubic and monoclinic zirconia, while the ZGYbY powder exhibited excellent stability due to the complete retention of the t′ zirconia phase and the transition from tetragonal to monoclinic phase upon cooling. These results make ZGYbY a promising material for TBC applications [13].

YiTao Wang and coworkers obtained by solid state reactions x mol% $ZrO_2$-$Gd_3NbO_7$ dense ceramics ($x$ = 0, 3, 6, 9, 12), a promising candidate for TBC applications. The $ZrO_2$-$Gd_3NbO_7$ ceramic showed only orthorhombic crystal structures with uniform grain size, ranging from 2 to 20 μm. The thermal conductivity of $ZrO_2$-$Gd_3NbO_7$ is much lower, 1.21–1.82 ($W·m^{-1}·K^{-1}$) from 25 to 900 °C, compared to $La_2Zr_2O_7$ (1.50–2.00 $W·m^{-1}·K^{-1}$), $Y_2SiO_5$, 8YSZ (2.50–3.00 $W·m^{-1}·K^{-1}$), $ZrO_2$-$Eu_3TaO_7$, $ZrO_2$-$DyTaO_4$, $ZrO_2$-$Y_2O_3$-$Ta_2O_5$ and YbO. The thermal expansion coefficients (TEC) for $ZrO_2$-$Gd_3NbO_7$ range from 9.71–10.60 × $10^{-6}$ $K^{-1}$ (1200 °C), being higher than those of $La_2Zr_2O_7$ (9.00 × $10^6$ $K^{-1}$), $La_2SiO_5$ (6.90–8.80 × $10^{-6}$ $K^{-1}$) and YSZ (10.00 × $10^{-6}$ $K^{-1}$). The mechanical properties of $Gd_3NbO_7$ change slightly with increasing $ZrO_2$ content, Vickers hardness was about 10 GPa, and Young's modulus was 173–195 GPa, which was lower than YSZ (240 GPa). A lower Young's modulus results in materials with lower stiffness [29].

Chao Chen et al. investigated the effect of $Sc_2O_3$ in YSZ ceramics on hot corrosion behavior in molten salts $Na_2SO_4$ + $V_2O_5$ (50/50 wt.%) at 1000 °C. By chemical coprecipitation method followed by calcination they obtained powders of $xSc_2O_3$–$1.5Y_2O_3$-$ZrO_2$ (x = 4.5, 5.5, 6.5, in mol%) and 4.5 mol% $Y_2O_3$-$ZrO_2$, and compared their properties. They noted that with increasing $Sc_2O_3$ content, corrosion resistance and phase stability also increase. On the surface of the YSZ ceramic, corroded bar-shaped configuration appears, while it turns into granular shapes with reduced amounts when higher $Sc_2O_3$ was used to dope YSZ. Introduction of $Sc_2O_3$ into YSZ can enhance the M-O bond strength and the substrate's ability to prohibit vanadate penetration due to the relatively shorter ionic radius of $Sc^{3+}$ than that of $Y^{3+}$ and $Zr^{4+}$. Furthermore, the least possibility to react $Sc_2O_3$ with $V_2O_5$, according to the Gibbs free energy calculation and the Lewis acid-base rule, promotes the phase stability of the ScYSZ ceramics. The excellent hot corrosion resistance can be achieved for those ScYSZ ceramics with higher $Sc_2O_3$ content, such as $6.5Sc_{1.5}YSZ$ [30].

$ZrO_2$-$YO_{1.5}$-$TaO_{2.5}$ powders were prepared by Pitek et al. by inverse co-precipitation of mixed solutions of precursor salts and showed improved phase stability, corrosion resistance by sulfate/vanadate melting and resistance at least comparable to that of the reference 7YSZ material. The 16.6%$YO_{1.5}$ + 16.6%$TaO_{2.5}$ stabilized zirconia composition is tetragonal, stable against phase partitioning up to at least 1500 °C, and insensitive to the tetragonal–monoclinic transformation upon thermal cycling. This material also showed only slight evidence of corrosion in S/V melts after 500 h, compared with extensive attack and spallation at much shorter times (50–100 h) observed in 7YSZ [31].

In their work, Fan and colleagues studied the reduction mechanism in thermal conductivity of a series of $Sc_2O_3$-$Y_2O_3$ co-stabilized tetragonal $ZrO_2$ ceramics, finding a thermal conductivity that is 20–28% lower than zirconia stabilized with 6–8 wt.% of Y [32].

Sun at al. studied the phase stability and thermo-physical properties of 7.5 mol% $Sc_2O_3$ and $Gd_2O_3$ co-doped $ZrO_2$ (ScGdSZ). They stated that after 150 h heat treatment at 1400 °C, there was no monoclinic phase in $Sc_2O_3$ and $Gd_2O_3$ co-doped $ZrO_2$, indicating high phase stability of ScGdSZ. The substitution of $Sc_2O_3$ with $Gd_2O_3$ resulted in increased fraction of cubic phase, however, t′ phase was still the dominant phase when the substitution amount was less than 2 mol%. They noticed that with increasing the $Gd_2O_3$ proportion the thermal conductivity of ScGdSZ gradually decreased, owing to the larger cation radius and heavier atomic mass of $Gd^{3+}$. 3.7 $Sc_2O_3$ and 3.7 $Gd_2O_3$ co-doped $ZrO_2$ (in mol%) had the lowest thermal conductivity, which was 20% lower than 7.5 ScSZ and 40% lower than 4.5 YSZ, respectively [33]. Table 3 summarizes the influence of the synthesis route on the structure and morphology of $ZrO_2$ doped with mixed rare earth oxides.

**Table 3.** The structural and morphological properties of the different materials based on $ZrO_2$ doped with rare earth oxides.

| Nr. | Material | Technique | Structural and Morphological Properties | Reference |
|---|---|---|---|---|
| 1 | 8 wt.% rare earth (Ce, Gd and Y) doped $ZrO_2$ films | sol-gel | monoclinic and tetragonal phases; doping with RE ions, the crystallite size is reduced (D = 6752 nm for $ZrO_2$ undoped, D = 4640 nm for doped ZrO2). | [20] |
| 2 | Ce doped $ZrO_2$, Dy doped $ZrO_2$ | coprecipitation | tetragonal and monoclinic phases; doping with RE ions, the crystallite size is reduced (D = 12 nm for $ZrO_2$ undoped, D = 9 nm for $ZrO_2$/Dy, D = 6 nm $ZrO_2$/Ce). | [21] |
| 3 | (Gd (1–9 mol%) doped $ZrO_2$ | solution combustion | D = 25–35 nm | [23] |
| 4 | 1 mol% $RE_2O_3$ (RE = La, Nd, Gd, Yb) | by co-precipitation | tetragonal phase | [25] |
| 5 | partial substitution of $Y_2O_3$ with the equivalent $Ln_2O_3$ (Ln = Nd, Sm, Gd) in $ZrO_2$-$Nb_2O_5$-$Y_2O_3$ ceramics | - | tetragonal phase | [25] |
| 6 | $ZrO_2$-doped with (8.3% mol $Ta_2O_5$ + 8.3% mol$Y_2O_3$ by solid-state reaction.-1 mol.% $Ln_2O_3$ substituted 1 mol% $Y_2O_3$ | solid-state reaction | decreasing grain growth | [27] |
| 7 | $ZrO_2$-x-$Ln_2O_3$ (where Ln = Sm, Yb) | colloidal chemical synthesis | no phase transformation occurs during heating | [28] |
| 8 | $ZrO_2$ 9.5$Y_2O_3$ 5.6$Yb_2O_3$ 5.2$Gd_2O_3$ | chemical precipitation | excellent stability of tetragonal phase (of the thermal cycle at 1300 °C for 50 h) | [13] |
| 9 | $x$ mol% $ZrO_2$-$Gd_3NbO_7$ dense ceramics ($x$ = 0, 3, 6, 9, 12) | solid state reactions | D = 2–20 μm | [29] |
| 10 | $x$$Sc_2O_3$−1.5$Y_2O_3$-$ZrO_2$ ($x$ = 4.5, 5.5, 6.5, in mol%) | chemical coprecipitation | increasing $Sc_2O_3$ content, phase stability also increases. | [30] |
| 11 | $ZrO_2$-$YO_{1.5}$-$TaO_{2.5}$ | co-precipitation | tetragonal phase | [31] |

Table 4 summarizes the solid phase expected to exist in $ZrO_2$-$Y_2O_3$-$Ln_2O_3$ systems in thermal equilibrium conditions, based on the analysis of the corresponding thermal phase diagrams, assessing the mutual compatibility between YSZ and corresponding zirconates.

**Table 4.** Equilibrium compounds in the ternary diagrams $ZrO_2$–$Y_2O_3$–$Ln_2O_3$ (where $Ln$ = $Gd_2O_3$, $Sm_2O_3$, $CeO_2$, $La_2O_3$, and $Nd_2O_3$).

| Material | T [°C] | Compounds Present in the Ternary Diagram | Reference |
|---|---|---|---|
| $ZrO_2$–$Y_2O_3$–$Gd_2O_3$ | 1200 | F = fluorite; C = bixbyite; T = tetragonal $ZrO_2$; P = pyrochlore $Gd_2Zr_2O_7$; $\delta$ = $Y_4Zr_3O_{12}$ | [34] |
| | 1400 | F = fluorite; C = bixbyite; T = tetragonal $ZrO_2$; P = pyrochlore $Gd_2Zr_2O_7$; $\beta$ = B-$Gd_2O_3$ | |
| | 1600 | F = fluorite; C = bixbyite; T = tetragonal $ZrO_2$; $\beta$ = B-$Gd_2O_3$ | |
| $ZrO_2$–$Y_2O_3$–$Sm_2O_3$ | 1250 | F = cubic (fluorite structure) $(Zr_{1-x}(Y,Sm)x)O_{2-0.5x}$ ss; T = tetragonal $(Zr_{1-x}(Y,Sm)_x)O_{2-0.5x}$ ss; Pyr = $(Sm,Y)_2Zr_2O_7$ss (pyrochlore); B = monoclinic $(Sm,Y)_2O_3$ ss; C = cubic $(Y,Sm)_2O_3$ ss. | [35] |
| | 1400 | | |
| | 1600 | | |
| $ZrO_2$–$Y_2O_3$–$CeO_2$ | 1250 | T = tetragonal; F = cubic fluorite; C = cubic; $\delta$ = $Zr_3(Y_{1-x}Ce_x)_4O_{12}$. | [36] |
| | 1400 | Tss = tetragonal solid solution; Fss = fluorite-type solid solution; Css = body-centered cubic solid solution. | [37] |
| | 1600 | Tss = tetragonal zirconia structure; Css = cubic fluorite structure; Yss = C-type yttria structure. | [38] |
| $ZrO_2$–$Y_2O_3$–$La_2O_3$ | 1250 | A = hexagonal A-$La_2O_3$; B = monoclinic phase; C = cubic phase; F = fluorite phase; LaYP = $La_2Y_2O_6$ pyrochlore phase; Pyr = La–Y–Zr phase. | [39] |
| | 1400 | A = hexagonal A-$La_2O_3$; B = monoclinic phase; C = cubic phase; F = fluorite phase; LaYP = $La_2Y_2O_6$ pyrochlore phase; Pyr = La–Y–Zr phase. | |
| | 1600 | A =hexagonal A-$La_2O_3$; B = monoclinic phase; C = cubic phase; F = fluorite phase; LaYP = $La_2Y_2O_6$ pyrochlore phase; Pyr = La–Y–Zr phase. | |
| $ZrO_2$–$Y_2O_3$–$Nd_2O_3$ | 1250 | A = RE hexagonal solid solution; B = RE monoclinic solid solution; C = RE cubic solid solution; F = cubic fluorite solid solution; T = trigonal solid solution; Pyr = $Nd_2Zr_2O_7$ pyrochlore. | [40] |
| | 1400 | | |
| | 1600 | | |

## 3. REO-Doped $ZrO_2$-Based Thermal Barrier Coatings

Coatings obtained from YSZ powder co-doped with Gd and Yb synthesized by hydrothermal method have also shown a higher thermal insulation capacity than the YSZ coatings [18]. YSZ ceramics co-doped with Gd and Yb produced by co-precipitation and calcination exhibit a better corrosion resistance than YSZ [10]. LZ/YSZ double-layer coating has a higher strength and oxidation resistance. LZ prevents the rapid oxidation of bondcoat and the rapid growth of thermal grown oxide (TGO). $La^{3+}$ reduces the residual stress because it can control the phase transformation of TGO, by reducing the transition rate from the metastable phase $\theta$-$Al_2O_3$ to the stable phase $\alpha$-$Al_2O_3$ [16].

Zirconia stabilized with $xSc_2O_3$–20 mol% $CeO_2$ ($x$ = 3.6, 4.78, 5.63, 6.31, and 8 mol%) were successfully synthesized and the resistance to thermal shock was studied. After heat treatment at 1400 °C for 25 h, the results showed that the optimum structure was obtained for $x$ = 4.78 due to avoding the formation of monoclinic and cubic phases in the structure of the stable tetragonal phase (t′) [41].

Erbia-doped YSZ coatings fabricated by electron beam evaporation develop thicker oxide layers upon oxidation that lower even further their thermal diffusivity (conductivity) [42].

Rare-earth zirconates with perovskite or pyrochlore structure are presently considered as emerging TBC materials for the future. These pyrochlore phases are stable up to the melting point (around 2300 °C), making those potential TBC materials for higher application temperatures. They also have a low intrinsic thermal conductivity, associated with the complexity of crystallographic structure and difference in the number and types of atoms in a unit cell. At 1000 °C, the thermal conductivities are 1.5–1.6 $W·m^{-1}·K^{-1}$ and 1.2–1.3 $W·m^{-1}·K^{-1}$ for dense $La_2Zr_2O_7$ (LZO) and $Nd_2Zr_2O_7$, respectively. At 700 °C, the thermal conductivity values range from 1.5 to 1.6 $W·m^{-1}·K^{-1}$ for dense $Gd_2Zr_2O_7$ (GZO), $Nd_2Zr_2O_7$ (NZO), and $Sm_2Zr_2O_7$ (SZO) compared to 2.3 $W·m^{-1}·K^{-1}$ for dense 7YSZ [5]. The low thermal conductivity property comes from the fact that its pyrochloral structure ($A_2B_2O_7$) produces large quantities of oxygen ion vacancies and from the ionic mass difference between $La^{3+}$ and $Zr^{4+}$ which results in the phonon scattering [42,43]. Previous work has indicated that LZO exists in a single stable phase of pyrochloride up to its melting point (~2300 °C). Although LZO is considered a promising candidate for application as thermal barrier coatings due to its properties, its fracture toughness (0.9–1.3 $MPa·m^{0.5}$) is lower than that of YSZ (~3 $MPa·m^{0.5}$) [44,45]. Also, the lower coefficient of thermal expansion (CTE) of LZO may increase the mismatch stress in TBCs system during thermal cycling [46,47]. Below are presented the results of several studies that considered the use of LZO as a material for TBC.

Five different TBC systems, YSZ, $La_2Zr_2O_7$, $Gd_2Zr_2O_7$, $YSZ/La_2Zr_2O_7$, and $YSZ/Gd_2Zr_2O_7$, were produced and exposed to furnace thermal cyclic oxidation tests. The deposition of protective ceramic top coats was performed with Electron-Beam Physical Vapor Deposition (EB–PVD), single-layer coatings having a thickness of approximately 200 μm, while the thickness of each two-layer coating was approximately 100 μm. It was found that TBC double layers have a higher lifetime compared to TBCs with single layer. As a result of the performed cycling tests, the best performance was exhibited by the $YSZ/Gd_2Zr_2O_7$ coating system whereas $La_2Zr_2O_7$ TBC system displayed the lowest performance [48].

To prove the effect of the $La_2Zr_2O_7$ layer deposited on the YSZ layer on the thermal conductivity and oxygen penetration, two systems were exposed to isothermal and thermal cyclic oxidation tests. The two systems, YSZ and $YSZ/La_2Zr_2O_7$ top coats were deposited using EB–PVD technique. The stress distribution showed that the YSZ/LZ system has a longer life span than the YSZ TBC. Although the oxidation and thermal cyclic tests showed that double layer YSZ/LZ TBC exhibited better performance than single layer YSZ TBC, there is no drastic difference between the thermal performances of TBCs [49].

Another study suggested the use of $La_2Zr_2O_7$ in the design of TBC, for its low thermal conductivity. Although YSZ has been used as a top coat for TBC due to its thermal diffusivity, the major disadvantage of its use is its limited long-term operating temperature. Thus, two systems of functional graded coatings were obtained based on $Al_2O_3/La_2Zr_2O_7/YSZ$ ceramics: a five-layer system ($Al_2O_3/75\%Al_2O_3$ + $25\%La_2Zr_2O_7/25\%Al_2O_3$ + $25\%YSZ$ + $50\%La_2Zr_2O_7/50\%YSZ$ + $50\%La_2Zr_2O_7/75\%YSZ$ + $25\%La_2Zr_2O_7$) and a six-layer system ($Al_2O_3/75\%Al_2O_3$ + $25\%La_2Zr_2O_7/25\%Al_2O_3$ + $25\%YSZ$ + $50\%La_2Zr_2O_7/50\%YSZ$ + $50\%La_2Zr_2O_7/75\%YSZ$ + $25\%La_2Zr_2O_7/100\%$ YSZ), using atmospheric plasma spray technique (APS). The researchers used alumina directly as a starting layer, near the metal substrate, because it acts as an oxygen diffusion barrier that protects the metal substrate from oxidation. $La_2Zr_2O_7$ powder was prepared using the sol-gel technique. It was found that samples coated with YSZ over 75% YSZ + 25% $La_2Zr_2O_7$ possess higher resistance to the attack of a $Na_2SO_4$ + $V_2O_5$ molten mixture up to 50 h of treatment than those coated with 75%YSZ + 25% $La_2Zr_2O_7$ top coats [50].

A comparison was made between YSZ as a conventional ceramic top coating material, $Gd_2Zr_2O$ and $YSZ/Gd_2Zr_2O$ as new generation coating materials with rare earth zirconate content. These materials were deposited as ceramic top coatings with EB–PVD method onto the CoNiCrAlY bond coat. These samples were subjected to hot corrosion tests by spreading mixtures of 55% $V_2O_5$ and 45% $Na_2SO_4$ salt at 5 h intervals at 1000 °C. It turned out that double coatings of $YSZ/Gd_2Zr_2O_7$ proved to be more resistant to hot corrosion than other single coatings [51].

$YSZ/Gd_2Zr_2O_7$ system was developed by the APS technique. Plasma grade flowable 8 wt.% YSZ and GZO powders were prepared by a single step co-precipitation technique. FE–SEM cross-section analysis of the YSZ/GZO bilayer coating after corrosion test demonstrated the effectiveness of the bilayer design in preventing the penetration of corrosive salts to the YSZ layer. Also, YSZ/GZO bilayer TBC exhibited a higher thermal cyclic life (300 cycles) than the single layer 8YSZ (175 cycles) coatings at 1100 °C [52].

Besides the above mentioned and commonly employed techniques for the fabrication of zirconia-based thermal barrier coatings, namely EB–PVD and plasma spraying (PS), also the sputtering technique is sometimes used to fabricated TBCs, especially in combination with the other techniques. Sputter-deposited films can provide good insulating properties, but not as high as coatings fabricated by EB–PVD or PS, due to the lower thickness of coatings deposited by sputtering (usually a few microns), so this technique is not commonly used alone for the deposition of commercial thermal barrier coatings. However, sputtered coatings can result interesting for example when good thermal insulting properties are desired together with strong corrosion resistance, thanks to the dense structure got in sputtered films and their strong adhesion aided by lower stresses deriving from the lower thickness [53].

For example, Amaya et al. [53] deposited 8 mol% Yttria-Stabilized Zirconia (8YSZ) by radio frequency (rf) magnetron sputtering as thermal barrier coating with also anticorrosion function on AISI-304 stainless steel. Prior to the YSZ growth, an $Al/Al_2O_3$ graded buffer layer was deposited to improve the adhesion of the subsequent coating. The corrosion tests were conducted at temperature up to 700 °C in air and at exposure time up to 6 h, showing significant differences (one to three orders of magnitude) of corrosion resistance between the coated and uncoated steel.

Rösemann et al. in refs. [54,55] and their other works cited therein, studied the influence of different process parameters on the properties of fully and partially yttria-stabilized zirconia (8 mol% FYSZ and 4 mol% PYSZ, respectively) coatings for potential use in thermal barrier applications. In particular, they used the gas flow sputtering (GFS) technique, allowing higher deposition rates with respect to conventional magnetron sputtering, thus depositing coatings with thicknesses up to some tens of microns. The influence of different process parameters (bias voltage, oxygen flow rate, substrate temperature) was evaluated on the morphological, compositional and crystallographic properties of the grown films, and thermal cycling experiments were conducted between 100 °C and 1050 °C. In particular, a strong effect of the substrate temperature on the coating properties was found.

In a work by Noor-A-Alam et al. [56], coatings based on mixed hafnia and zirconia stabilized by yttria ($Y_2O_3$-stabilized $HfO_2$-$ZrO_2$, briefly YSHZ) were tested for TBC applications. Different coating compositions were studied by varying the relative concentration of $HfO_2$ and $ZrO_2$, at fixed $Y_2O_3$ stabilizer content of 7.5 mol%. The thermal stability of the YSHZ coatings was evaluated up to 1300 °C by XRD, SEM, and EDS analyses, showing no significant modifications of crystal structure, morphology and composition.

The use of other REOs like $Gd_2O_3$ as stabilizer in sputtered coatings based on zirconia with and without yttria was tested by Portinha et al. [57] as potential materials for TBCs. $ZrO_2Gd_2O_3$ and $ZrO_2Y_2O_3Gd_2O_3$ coatings were grown by DC magnetron sputtering at different compositions, with $Gd_2O_3$ content varying in the range from 6 to 38 wt.% in $Gd_2O_3$-stabilized $ZrO_2$, and $Y_2O_3$ and $Gd_2O_3$ contents in the respective ranges from 4 to 10 wt.% and from 11 to 20 wt.% in zirconia samples with both stabilizers. The zirconia phase was found to shift from tetragonal at lower stabilizers concentrations to cubic at increasing concentrations, while no significant changes were noticed in the coating morphology.

A summary of the sputtering parameters used in the works mentioned above are reported in Table 5.

**Table 5.** Deposition parameters, materials and film thicknesses for the coatings deposited by sputtering in the different works described in the text.

| Coating Material | Deposition Technique | Coating Thickness | Target | Power | Process Pressure | Substrate Temperature | Bias Voltage | Reference |
|---|---|---|---|---|---|---|---|---|
| 8YSZ | rf magnetron sputtering | ~2.5 μm | YSZ | 350 W | 0.7 Pa (Ar) | 300 °C | −20 V | [53] |
| 8 mol% and 4 mol% YSZ | gas flow sputtering | ~15–70 μm | ZrY alloy | 5 kW (DC) | 20–50 Pa (Ar + $O_2$) | 500–800 °C | 0—100 V (pulsed DC) | [54,55] and works cited therein |
| YSHZ | sputtering | ~1 μm | YSHZ | 100 W | /(Ar) | 500 °C | / | [56] |
| $Gd_2O_3$-stabilized $ZrO_2$ and $Gd_2O_3$-costabilized YSZ | DC magnetron sputtering | ~1.2–3.8 μm | Zr doped with Gd and/or Y | Voltage 260–290 V | 0.6 Pa (Ar + $O_2$) | 250 °C | −50 V (DC) | [57] |

In order to combine the strong adhesion characteristics of sputtered coatings together with the better thermal barrier properties induced by the thicker films deposited by plasma spray, the two techniques can be combined together in a multilayer structure, with the aim to reduce TBC failure due to coating spalling. To this aim, in [58] by Andritschky et al. a preliminary bond coat of yttria-partially-stabilized zirconia (PSZ, 8 μm thick) was deposited on the substrate by DC reactive magnetron sputtering, followed by a thicker PSZ coating (300 μm) deposited by plasma spray. Different thermal cycling tests were conducted on the analyzed coatings, showing no significant damage on the combined coatings under the experimental conditions used in the work.

A similar approach was used by Yao et al. [59], where a multilayered TBC was obtained by the combination of a bond coat layer deposited by sputtering and a top layer fabricated by EB–PVD, with the aim to improve the TBC performance thanks to reduced oxygen diffusion, inhibition of cracks propagation and stress relaxation, aided by the bond coat. In particular, a composite bond coat made of

$Y_2O_3$-doped $Al_2O_3$ and noble metals (Pt and Au) was grown on the substrate by depositing different alternating layers of $Y_2O_3$-doped $Al_2O_3$ (through rf magnetron sputtering from an $Al_2O_3$ target doped with 2 wt.% of $Y_2O_3$) and noble metal layers (through DC magnetron sputtering from a metal target), followed by thermal annealing, with a total thickness of about 1.4 μm. The subsequent 8YSZ top coat with thickness around 100 μm was then deposited on the bond coat by the EB–PVD process, and the combined coating was tested by high temperature cyclic oxidation tests at 1100 °C in air for 200 h. This composite coating showed strong adhesion on the substrate and resulted in a largely improved resistance to oxidation and spallation.

ZrYGdO coatings fabricated by cosputtering Y-and Gd-doped $ZrO_2$ targets were used as external barrier coatings on ion-plasma thermal barrier NiCrAlY (Re, Ta, Hf) + AlNiY (Hf) alloys and submitted to isothermal and cyclic heat-resistance tests. The effect of TBCs on the long-term strength at a test temperature of 1000 °C and on the high-cycle fatigue at 900 °C was investigated [60].

## 4. Perspectives for Using Zirconia Doped with Multicomponent Mixed REOs for TBC: Preliminary Results

From the critical analysis of the studies reported in the literature it is evident how the thermal properties of zirconia may be improved by using different rare earth oxides. The aim of the present work is to show the potential of using mixed rare earth oxides (REOs) as naturally occurring from monazite concentrates for obtaining stabilized $ZrO_2$ ceramic coatings for their potential used as TBCs.

Rare Earth Oxides naturally coexist in different ratios in concentrates such as monazite or bastnäsite. Their extraction, separation and purification in individual REOs is a process requiring special skills, hundreds of hydrometallurgical processing steps using organic solvent extraction, precipitation and calcination, with high environmental impact and energy consumption reflected in corresponding high prices. Despite the more and more theoretical and practical interest for using mixed REOs as dopants in high tech ceramics, only few attempts have been done to study the possibility to extract the naturally complex of REOs from concentrates and use them as multicomponent mixed doping oxides [61]. Here we present some preliminary results on the development of zirconia doped with multicomponent REOs simulating the composition from selected monazite concentrates after removal the radioactive elements (Th, U, and Ra). The aim is to further assess the potential use of naturally mixed REOs obtained directly from monazite concentrates.

Materials synthesis: In order to study the microstructure of mixed REOs-doped zirconia for potential applications in TBCs, five compositions of doped $ZrO_2$ powders have been obtained. In sample $ZrO_2$-RE1 the 8 wt.% $Y_2O_3$ usually employed as dopant for TBC was replaced with 8% synthetic REOs mixture with a ratio corresponding to natural occurrence in selected La-rich monazite concentrates. The other samples, denoted MxZy8La, MxZy8Sm, MxZy8Nd, and MxZy8Gd, represent $ZrO_2$ doped with 8 wt.% of each single element (La, Sm, Nd, and Gd, respectively), and were used as reference samples for the XRD analyses with the purpose to understand the XRD pattern of the REOs mixed co-doped powders.

All samples were prepared by a hydrothermal method at moderate temperatures (max. 250 °C) and pressures (max. 40 atm.). This method is characterized by three important advantages: improvement of the chemical reactivity, low energy consumption due to elimination of thermal treatment for crystallization, high homogeneity and control of nucleation and growth [62]. The synthesis was done from high purity raw materials: $Y_2O_3 > 99\%$-Merck, $La_2O_3 \geq 99.9\%$-Roth, $Nd_2O_3 \geq 99.9\%$-Alfa Aesar, $Sm_2O_3 \geq 99.9\%$-Alfa Aesar, $Gd_2O_3 \geq 99.9\%$-Alfa Aesar and $Yb_2O_3$ 99.9%-Alfa Aesar. Zirconium tetrachloride ($ZrCl_4$ 99% Merck) was used as raw material for preparing an aqueous Zr (IV) stock solution with programmed Zr concentration. The dissolution of REOs precursors in $ZrCl_4$ solution was done under vigorous mechanical stirring until a homogenous clear solution was obtained. Ammonia solution ($NH_3$ 25% p.a., Chimreactiv srl) agent was added as mineralizing until an alkaline suspension with pH~9 was obtained. The doped powder was obtained by hydrothermal treatment in an autoclave (Berghof, Germany, TEFLON-linen, 5-L capacity, maximum operation temperature 250 °C, maximum operating pressure 200 atm., endowed with water cooling system). The hydrothermal method used was described in [63]. The synthesis flowsheet is presented in Figure 1.

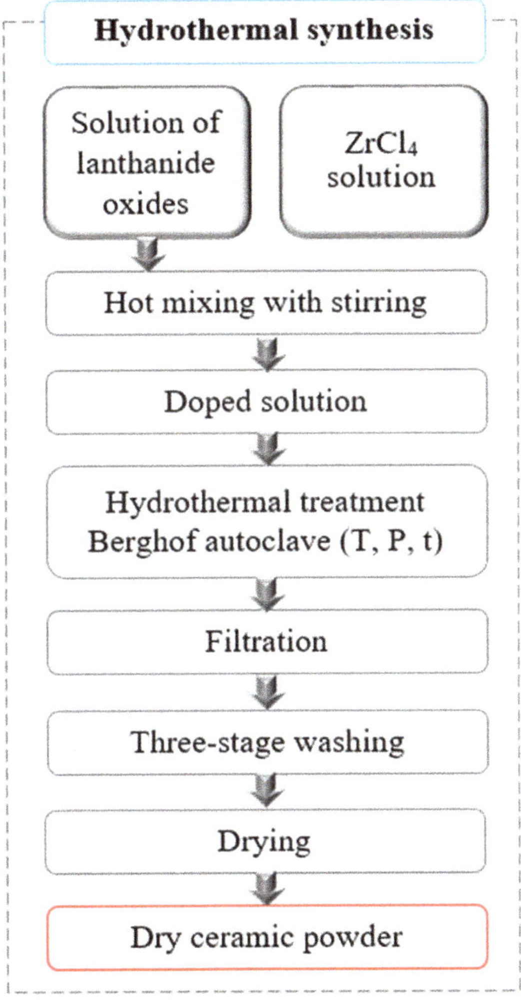


**Figure 1.** Schematic flowsheet of the hydrothermal process used for obtaining ZrO$_2$ powders doped with mixed rare earth oxides (REOs).

Spark plasma sintering (SPS) was used to achieve compacted materials virtually without porosity or bearing a minimal and almost insignificant porosity level [64]. All experiments were performed under a vacuum of 16.2 Pa with the pulse (3.3 ms) sequence for an applied voltage of 12:2 V (i.e., 12 ON/2 OFF). The samples were sintered at temperature of 1250 °C, for a dwell time of 20 min with heating and cooling rates of 10 °C/min. A known volume of 0.75 g of powder was used for each experiment. The experiment was carried out in a graphite mold with an inner diameter of 10 mm and an external diameter of 25 mm with internal diameter of the graphite die covered by flexible carbon foil (Papyex). The mold was covered with carbon fiber felt to limit the loss of heat radiation. The sintered pellets were used as source material for Electron-Beam Physical Vapor Deposition process.

Obtaining of mixed REOs-doped zirconia coatings. TB coatings were developed in a special designed EB–PVD thin film coating equipment (Torr International Inc., New Windsor, NY, USA) endowed with quartz sensors (QCM) mounted next to each crucible for monitoring the deposition rate and a software allowing the creation of complex deposition recipe that can be performed in automatic mode keeping constant the evaporation rate. The system is designed to combine multiple coatings, having 4 EB 10 KW power guns, each being equipped with four crucibles with body movement, allowing the continuous storage of 16 materials. The substrate material used was a high temperature NIMONIC 80A alloy in the form of sheets with sizes 30 mm × 50 mm, mounted on a heated support rotating at 20 rpm during coating process. All successive layers were deposited by EB–PVD process. Commercial NiCrAlY powders (Amperit) were used to deposit the bond coat prior to deposit the ceramic layer. Commercially $La_2Zr_2O_7$ (LZO) granulated powders (grain sizes 30–120 μm) and $Gd_2Zr_2O_7$ (GZO) granulated powders (grain sizes 45–140 μm) from Trans-Tech Ceramics and Advanced Materials USA were also used to deposit the outer layers, aiming to improve the thermal shock properties of the system. The schematic flowsheet of the whole process for obtaining mixed REOs doped $ZrO_2$ coatings is presented in Figure 2. The experimental parameters used for coating of sintered REOs-doped zirconia pellets are presented in Table 6.

**Table 6.** Working conditions and parameters.

| Material Used and Number of Crucibles | Substrate Heating Temperature (°C) | Start Vacuum (Pa) | Working Vacuum (Pa) | Deposition Rate (Å/s) | Total Deposition Time (h) | Maximum Power (KW) |
|---|---|---|---|---|---|---|
| 6 crucibles: 1xNiCrAlY 4 × ceramic pellets 1 × LZO, 1× GZO | 650 | $1.33 \times 10^{-4}$ | $5.33 \times 10^{-3}$ | 0.8–1.4 | 57 | 10 |

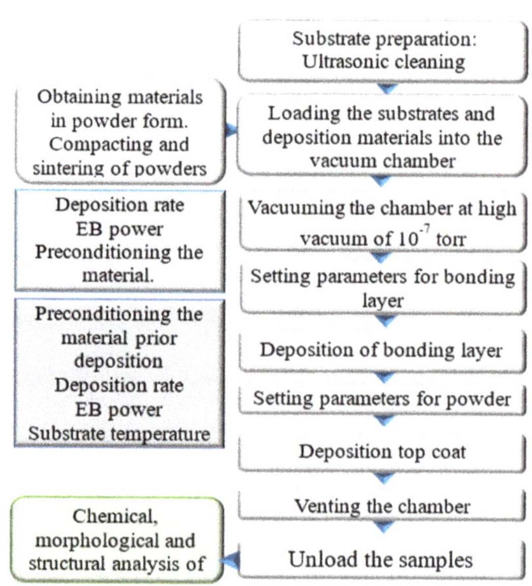


**Figure 2.** Schematic flowsheet for obtaining of mixed REO-doped $ZrO_2$ coatings.

Characterization methods: Chemical composition of REO-doped $ZrO_2$ was analyzed by Inductively Coupled Plasma-Optical Emission Spectrometry (Agilent 725 ICP-OES), according to ASTM E 1479-99(2011) standard. The microstructure of the as-obtained REO doped $ZrO_2$ powder was examined using a BRUKER D8 ADVANCE X-ray diffractometer (Bruker AXS Company, Germany) with monochromatic Cu Kα radiation, Bragg–Brentano diffraction method. Scans were obtained in the 2θ range 4–74° with a step size of 0.02° every 2.5 or 6 s. For the identification of the phases contained in the samples, data processing was performed using software package DIFFRAC.SUITE.EVA release 2016 by Bruker AXS Company, Karlsruhe, Germany; SLEVE + 2020 and ICDD PDF-4 + 2020 database edited by International Centre for Diffraction Data (ICDD). X-ray diffraction analysis on the deposited thin films was performed using an Empyrean PANalytrical equipment in Bragg–Brentano geometry equipped with an in-focus Cu (λ Cu Kα = 1.541874 Å) X-ray tube, hybrid monochromator on the incident side and parallel plate collimator mounted on the detector a PIXcel3D on the diffracted side. The spectra were acquired in the 2θ range of 20–80°, with acquisition step of 0.02° and acquisition time per step of 4s. X-ray diffraction spectra have been processed in the specialized software HighScore Plus 3.0e software (PANalytical, Almelo, The Netherlands).

The morphology of the powder samples was investigated by scanning electron microscopy (SEM) using a high-resolution microscope Quanta 250 (FEI Company, Eindhoven, The Netherlands), incorporated with Energy Dispersive X-Ray Spectrometer, produced by EDAX (Mahwah, NJ, US), consisting of ELEMENT Silicon Drift Detector Fixed, Element EDS Analysis Software Suite APEX™ 1.0, EDAX, Mahwah, NJ, USA.

The characterization of the films by SEM-EDS was performed using the QUANTA INSPECT F50 SEM (FEI Company, Eindhoven, The Netherlands) equipped with field emission gun electron (FEG) with 1.2 nm resolution and energy dispersive X-ray spectrometer (EDS), with the resolution at MnK of 133 eV.

Roughness tests were developed using a digital roughness meter, INSIZE-ISR-C002 (type: Inductive; probe head material: diamond; measuring force: 4 mN; units: μm/min; number of interruptions: 1 to 5; crossing speed: 0.5 mms$^{-1}$ and mms$^{-1}$; resolution (Ra): 0.001 μm).

The thermal shock test was performed with the help of a Quick Thermal Shock Installation (QTS-INCAS Bucharest, Romania), which has a maximum oven operating temperature of 1750 °C. The specimens were inserted into the oven by the action of a robot axis and kept in the oven for 5 min. The cooling stage is 60 s to reach the room temperature, which is produced by a jet of compressed air over a 5 mm nozzle at a pressure of 8 bar ± 10% at a distance of 45 mm from the sample under 45° degree of inclination. Temperature data are collected using 1 Pt/PtRh thermocouple inside the oven and 2 pyrometers outside the oven, for the temperature range −40–1750 °C. A pyrometer mounted on the oven monitors the temperature variation on the test tube inside the oven.

The chemical analysis of powders synthesized in hydrothermal conditions is presented in Table 7 and it is in accordance with the designed compositions. The XRD patterns of samples are presented in Figure 3. The quantitative phase analysis is presented in Table 8.

**Table 7.** Chemical composition of REO-doped $ZrO_2$ powders.

| Sample | | Chemical Analysis | | | | | | |
|---|---|---|---|---|---|---|---|---|
| | | La | Gd | Y | Yb | Sm | Nd | Zr |
| $ZrO_2$-RE1 | wt.% | 3.49 | 0.278 | 0.46 | 0.0032 | 0.409 | 2.33 | 52.19 |
| MxZy8La | wt.% | 8 | - | 5.50 | - | <0.004 | - | 63.32 |
| MxZy8Sm | wt.% | - | - | 5.77 | - | 9.28 | | 60.46 |
| MxZy8Nd | wt.% | - | - | 5.70 | - | - | 8.09 | 61.50 |
| MxZy8Gd | wt.% | - | 9.93 | 5.68 | - | - | - | 60.03 |

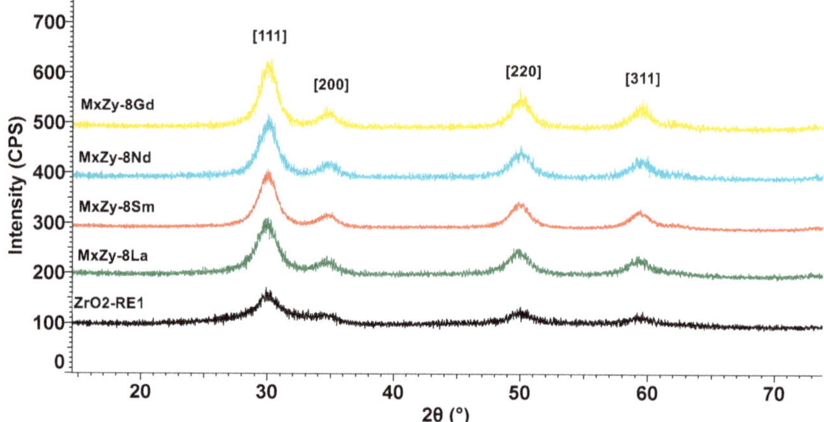

**Figure 3.** Diffraction spectrum for ZrO$_2$-RE1, MxZy8La, MxZy8Sm, MxZy8Nd, and MxZy8Gd samples.

**Table 8.** Quantitative phase analysis of ZrO$_2$ doped powders.

| Sample | Major Phases | Formula | PDF File | Crystallization System |
|---|---|---|---|---|
| ZrO$_2$-RE1 | Zirconium Yttrium Oxide | Solid solution | PDF 00-060-0505 | Tetragonal |
| MxZy8La | Zirconium Yttrium Oxide type | Solid solution | PDF-01-077-2286 | Cubic |
|  | Baddeleyite | ZrO$_2$ | PDF-00-036-0420 | Monoclinic |
| MxZy8Sm | Zirconium Yttrium Oxide type | Solid solution | PDF 01-077-2286 | Cubic |
| MxZy8Nd | Zirconium Yttrium Oxide type | Solid solution | PDF 01-077-2286 | Cubic |
| MxZy8Gd | Zirconium Yttrium Oxide type | Solid solution | PDF 01-077-2286 | Cubic |
|  | Baddeleyite | ZrO$_2$ | PDF 00-036-0420 | Monoclinic |

It may be observed that ZrO$_2$ doped with 8 wt.% La$_2$O$_3$, Sm$_2$O$_3$, and Nd$_2$O$_3$ are monophasic cubic solid solution, while ZrO$_2$ doped with 8 wt.% Gd$_2$O$_3$ consists of cubic solid solution as major phase and monoclinic ZrO$_2$ as minor phase. Sample ZrO$_2$-RE1 (composition simulating La-rich monazite concentrates) corresponds to a monophasic tetragonal ZrO$_2$ solid solution.

The morphology of these powders is presented in Figure 4. All hydrothermally synthesized powders consist of granular aggregates with dimensions up to tens of microns. EDS analysis (not shown) confirms the presence of REOs doping elements uniformly distributed in the aggregates.

Different coatings architectures were obtained from ZrO$_2$-RE1 pellets, LZO and GZO granulated powders. The section view of a typical architecture of EB–PVD coatings on Nimonic 80 A substrate (Figure 5) highlights the presence of the 4 layers arranged as follows: NiCrAlY bonding layer with 0.556 $\pm$ 0.072 μm thickness, YSZ with 1.984 $\pm$ 0.151 μm layer thickness, LZO layer thickness 4.245 $\pm$ 0.119 μm and GZO with a layer thickness of 4.618 $\pm$ 0.191 μm. The total thickness of the deposit is around 11.50 μm. Elemental mapping, through the colors chosen for the specific elements, confirms the clear limits of each layer.

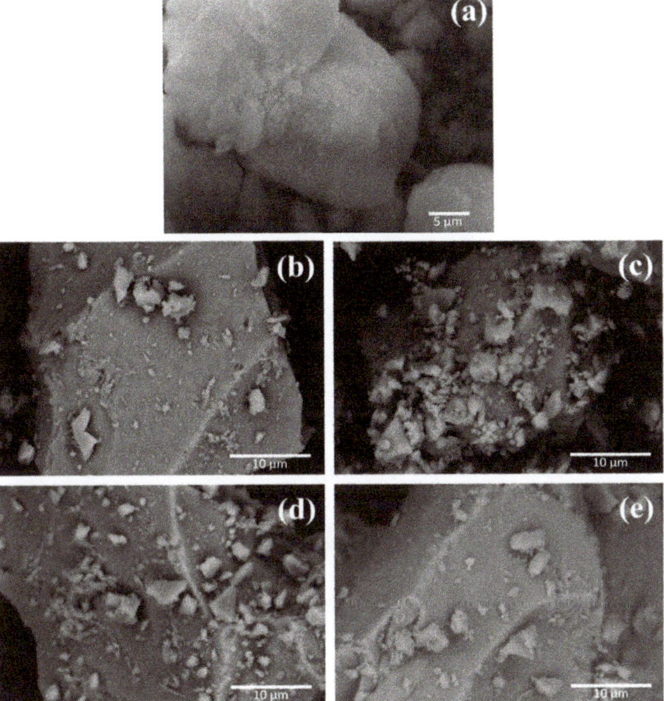

**Figure 4.** Scanning electron microscopy (SEM Quanta 250 FEI Company, Eindhoven, The Netherlands) images for samples: (**a**) ZrO2-RE1; (**b**) MxZy8La; (**c**) MxZy8Sm; (**d**) MxZy8Nd; and (**e**) MxZy8Gd.

SEM analysis on the coated surface (Figure 5a) shows that the deposition is uniform with a continuous film on the substrate. Polyhedral grains with well-defined edges and dimensions between 190 and 380 nm are identified.

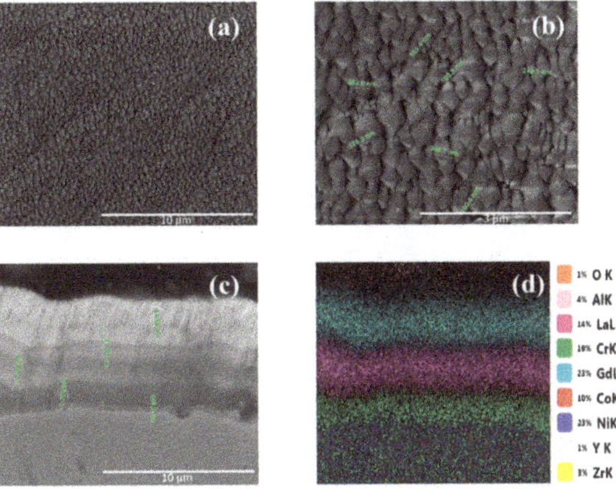


**Figure 5.** SEM images (50,000×), QUANTA INSPECT F50, FEI Company, Eindhoven, The Netherlands, representing: (**a**) and (**b**) the surface of the NiCrAlY/YSZ/LZO/GZO coating deposited on Nimonic substrate at different magnifications; (**c**) the section view of NiCrAlY/ZrO2-RE1/LZO/GZO type deposition on Nimonic substrate, and (**d**) the elemental mapping.

In Figure 6, the XRD spectrum of the Nimonic 80 A coated with four layers of material (NiCrAlY/YSZ/LZO/GZO) is presented, showing the presence of $Gd_2Zr_2O_7$ (ICDD PDF4 + 01-078-4083) and $La_2Zr_2O_7$ (ICDD PDF4 + 00-050-0837). The presence of REOs doped $ZrO_2$ solid solution cannot be observed, due to the attenuation of the X-ray radiation by the upper layers.

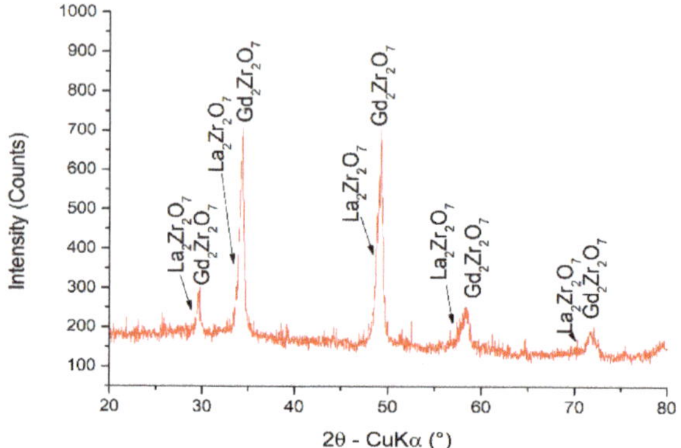

**Figure 6.** X-ray diffraction spectrum obtained for Nimonic with 4 layers of material (NiCrAlY/ZrO2-RE1/LZO/GZO).

The Nimonic sample coated with four layers of NiCrAlY/$ZrO_2$-RE1/LZO/GZO material was subjected to heat treatment at a temperature of 1250 °C for 10 min in nitrogen to determine its behavior under high temperature conditions. The SEM micrographs obtained after treatment are shown in Figure 7. One can observe in Figure 7b and from the corresponding map (Figure 7d), that during heat treatment, most probably during cooling in air, the substrate/bondcoat became oxidized, showing typical oxide scale morphology between substrate and ceramic coating and enrichment of Cr, most probably with the formation of $Cr_2O_3$ oxide. Small cracks appear at the intergranular boundaries in the $ZrO_2$-REO1 layer and the diffusion of Ti and Cr from the substrate towards the cracking layer, as well as the formation of $TiO_2$ were observed.

The corresponding XRD spectrum of the coating after the heat treatment (Figure 8) shows the presence of $Gd_2Zr_2O_7$ (ICDD PDF4 + 01-078-4083) and $La_2Zr_2O_7$ (ICDD PDF4 + 00-050-0837) phases, while the presence of $ZrO_2$-RE1 and other secondary phases formed at the substrate/ceramic layer interface is hindered.

From the roughness investigations results presented in Figure 9, a slight change in the mean surface roughness from 0.448 μm of the Nimonic 80A substrate to 0.521 μm for the coated Nimonic substrate may be observed. These values are very similar, thus proving that the coatings grow uniformly following the substrate morphology.

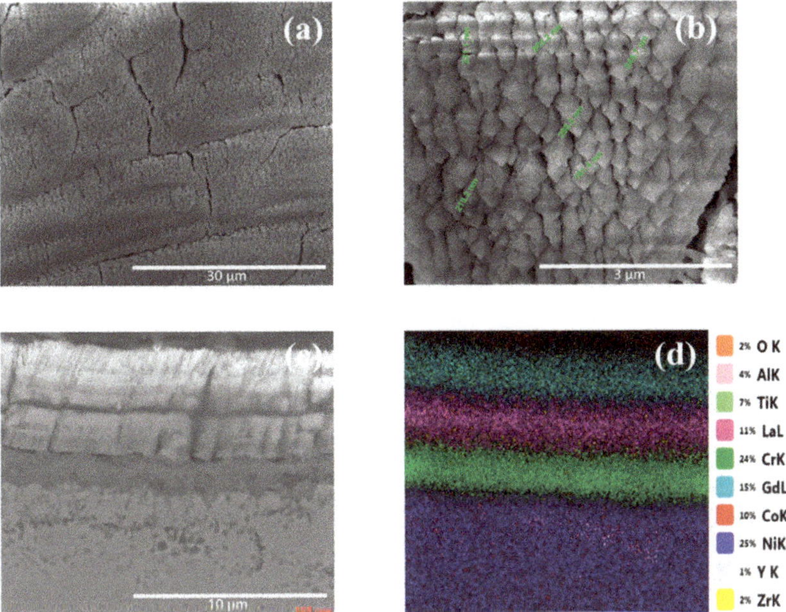

**Figure 7.** SEM images (QUANTA INSPECT F50, FEI Company, Eindhoven, The Netherlands) representing (**a**) and (**b**) the surface of the NiCrAlY/ZrO2-RE1/LZO/GZO multilayer coating on the Nimonic substrate after heat treatment at 1250 °C (40,000×); (**c**) cross section of the NiCrAlY/ZrO2-RE1/LZO/GZO on Nimonic substrate; and (**d**) elemental mapping after heat treatment.

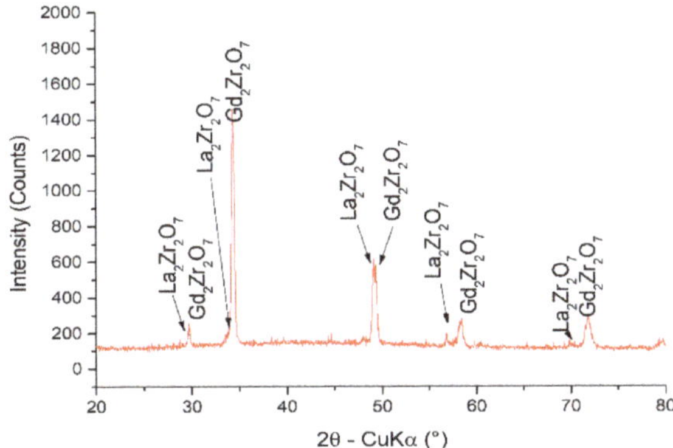

**Figure 8.** X-ray diffraction spectrum obtained for Nimonic with 4 layers of material (NiCrAlY/YSZ/LZO/GZO) after heat treatment at 1250 °C.

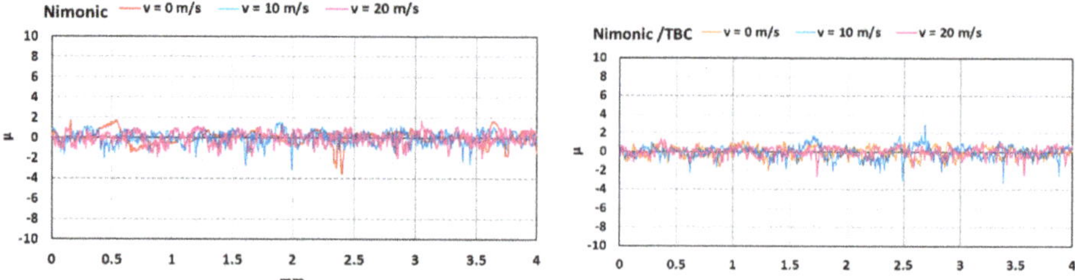



**Figure 9.** Surface roughness investigations of uncoated and coated Nimonic 80A substrate.

The results of thermal shock tests are presented in Figures 10 and 11. This test shows the number of heating and cooling cycles that a coated material may support without delamination of the coating layers. Comparison between thermal shock values of new coatings architecture with existing ones may be a method to assess their potential application in aeronautics. The results show a satisfactory behavior obtained after a number of minimum 150 cycles in the temperature ranges 1200–1300 °C for the proposed NiCrAlY/ZrO2-RE1/LZO/GZO coating architecture with a total thickness of about 11.5 μm. The results are comparable with those of traditional YSZ coatings with thickness greater than 100 μm [65], showing the ability of the new coating materials to further improve the thermal properties of TBCs for aeronautical applications at much lower thickness.

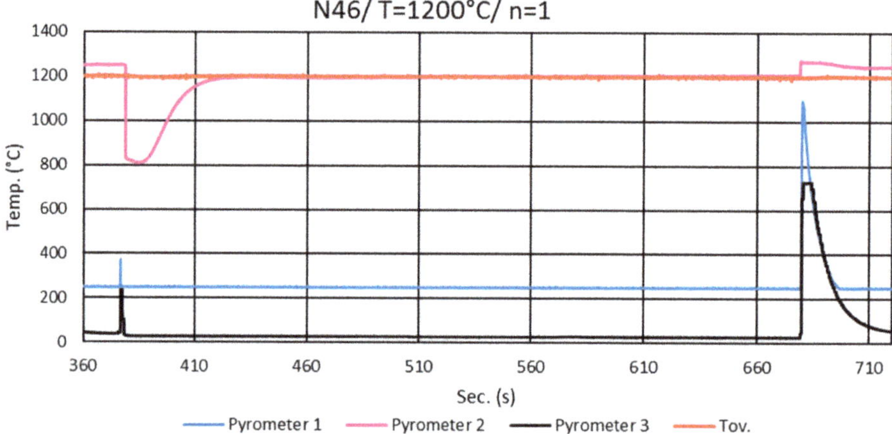



**Figure 10.** Thermal shock tests. The red line (TCup)—represents the recorded temperature of the oven by the Pt/PtRh thermocouple mounted inside the oven; magenta line (Pyrometer 2) represents the temperature of the sample over the heating stage during the test, which shows a peak starting from low temperature and a stabilized temperature for dwell time. The blue and black lines represent the temperature change of the sample over the cooling stage by Pyrometer 1 and Pyrometer 3, respectively.

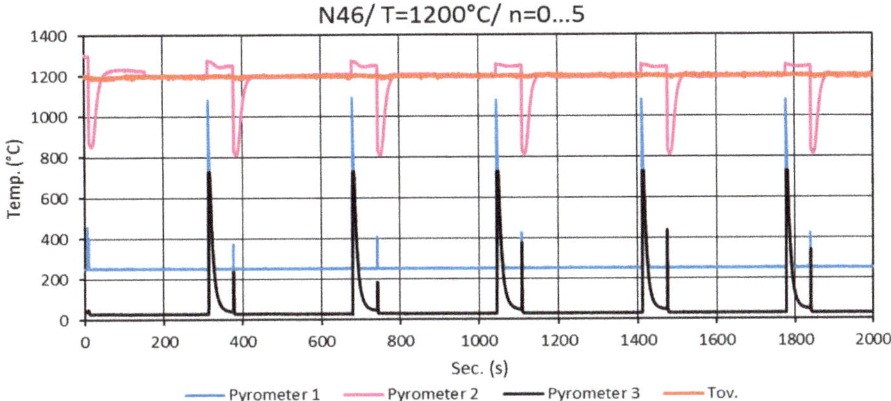



**Figure 11.** Thermogram for thermal shock cycles 38–43, temperature range 800–1300 °C.

## 5. Conclusions and Future Prospects

During recent years, it was demonstrated that co-doping of zirconia ceramics with REOs may avoid grain size coarsening due to interface segregation, enhancing its ionic conductivity and sinterability. $ZrO_2$ doped with 7–8 wt.% $Y_2O_3$ (8YSZ) are presently considered as a "gold standard" in coating materials for TBC applications. New TBCs that can operate at a temperature above 1200 °C are needed and doping $ZrO_2$ ceramics with mixed rare earth oxides is a possible solution due to the high temperature transformation and the low thermal conductivity. The influence of the synthesis route on the microstructure and properties of $ZrO_2$ doped with different REOs mixed compositions are presented and discussed.

Rare-earth zirconates with perovskite or pyrochlore structure are also considered as emerging TBC materials for the future, due to their stability up to the melting point, low intrinsic thermal conductivity, associated with the complexity of crystallographic structure and large quantities of oxygen ion vacancies.

Different physical vapor deposition methods have been proposed to obtain TBCs from these high temperature oxides. EB–PVD method was largely used to obtain coatings with desired properties for aeronautics and energy generation systems due to the specific columnar growth and good adhesion to the substrate.

In this paper, we presented some preliminary results on the development of zirconia doped with multicomponent REOs simulating the composition existing in some selected monazite concentrates after removal of the radioactive elements (Th, U, and Ra). The aim is to further assess the potential use of naturally mixed REOs obtained directly from monazite concentrates. $ZrO_2$ powders doped with 8% synthetic REOs mixture (containing La, Gd, Y, Nd, Gd, and Yb with a ratio corresponding to natural occurrence in selected La-rich monazite concentrates) have been prepared by a hydrothermal method at moderate temperatures (max. 250 °C) and pressures (max. 40 atm.). The XRD pattern of this powder corresponds to a monophasic tetragonal $ZrO_2$ solid solution, consisting of granular aggregates with dimensions up to tens of microns. EDS analysis confirms the presence of REOs doping elements uniformly distributed in the aggregates.

Coatings architectures deposited on Nimonic 80 A substrates were further obtained using these powders, consisting of NiCrAlY bonding layer with 556 nm thickness, YSZ with 1.98 μm layer thickness, LZO layer thickness 4.25 μm and GZO with a layer thickness of 4.62 μm. The results of thermal shock tests show a satisfactory behavior for a number of minimum 150 cycles in the temperature ranges 1200–1300 °C for the proposed coating architecture with a total thickness around 11.5 μm. The results are comparable with those of traditional YSZ coatings with thickness >100 μm, showing the ability of the new coating materials to further improve the thermal properties of TBCs for aeronautical applications.

Works are under progress to obtain TBC coatings based on $ZrO_2$ doped with synthetic REOs mixtures simulating Ce-rich monazite concentrates. The results will be further compared to those obtained using natural mixed REOs obtained from monazite concentrates, to assess the possibility to use them as dopants and demonstrate their economic impact in aeronautics and energy generation.

**Author Contributions:** Conceptualization, R.R.P.; methodology, M.C. and M.P.; validation, A.M.M. and V.D.; formal analysis, B.S.V.; investigation, M.B., A.V.S. and R.T.; resources, S.V.; data curation, A.E.S.; writing—original draft preparation, A.E.S., D.V., M.L.G. and R.R.P.; writing—review and editing, D.V. and M.L.G.; supervision, R.R.P. and M.L.G. All authors have read and agreed to the published version of the manuscript.

**Funding:** This research was funded by H2020 ERAMIN II Programme, MONAMIX project ID 87, financed in the frame of grant 50/2018 UEFISCDI Romania, Ctr. ANR-17-MIN2-0003-03 France and Ctr. MIUR 8361/05 May 2017-CUP n. B86G17000750001 Italy. Mircea Corban acknowledges the funding in the frame of project PN 19 19 04 01 financed by Romanian Ministry for Education and Research. Mihai Botan akcnowledges the funding by the European Social Fund through the Sectoral Operational Programme Human Capital 2014–2020, through the Financial Agreement with the title "Scholarships for entrepreneurial education among doctoral students and postdoctoral researchers", Acronym Be Entrepreneur!, Contract no. 51680/09.07.2019-SMIS code: 124539.

**Acknowledgments:** Radu R. Piticescu, Maria Luisa Grilli, Daniele Valerini and Mythili Prakasam acknowledge also the networking support from COST Action CA 15102 CRM Extreme (2016–2020) and CIG–15102 ITHACA (2020–2021).

**Conflicts of Interest:** The authors declare no conflict of interest.

## References

1. Darolia, R. Thermal barrier coatings technology: Critical review, progress update, remaining challenges and prospects. *Int. Mater. Rev.* **2013**, *58*, 315–348. [CrossRef]

2. Castro, J.; Nafsin, N.J. Direct measurements of quasi-zero grain boundary energies in Ceramics. *J. Mater. Res.* **2017**, *32*, 166–173.

3. Dragut, D.V.; Badilita, V.; Motoc, A.M.; Piticescu, R.R.; Zhao, J.; Hijji, H.; Conte, L. Thermal stability and field assisted sintering of cerium-doped YSZ ceramic nanoparticles obtained via a hydrothermal process. *Manuf. Rev.* **2017**, *4*, 11. [CrossRef]

4. Sobetkii, A.; Rinaldi, A.; Cuesta-Lopez, S.; Prakasam, M.; Largeteau, A.; Plaiasu, G.; Piticescu, R.R. Novel high temperature oxide ceramic coatings: Synthesis, properties and applications. In Proceedings of the International Conference FiMPART, Bordeaux, France, 9–12 July 2017.

5. Ekström, M.; Thibblin, A.; Tjernberg, A.; Blomqvist, C.; Jonsson, S. Evaluation of internal thermal barrier coatings for exhaust manifolds. *Surf. Coat. Technol.* **2015**, *272*, 198–212. [CrossRef]

6. Patnaik, P.; Huang, X.; Singh, J. State of the Art and Future Trends in the Development of Thermal Barrier Coating Systems. In *Innovative Missile Systems*; Meeting Proceedings RTO-MP-AVT–135, Paper 38; RTO: Neuilly-sur-Seine, France, 2006; Volume 38, p. 20.

7. Boke, Y.E.; Altun, O. Heat transfer analysis of thermal barrier coatings on a metal substrate. *J. Therm. Sci. Technol.* **2013**, *33*, 101–109.

8. Mehdi, K.; Soheil, N. The evolution of fracture process zones in as-received and oxidized air. *Surf. Coat. Technol.* **2019**, *377*, 124885.

9. Kadir, M.D.; Yasin, O.; Hayrettin, A.; Abdullah, C.K. Evaluation of oxidation and thermal cyclic behavior of YSZ. Gd2Zr2O7 and YSZ/Gd2Zr2O7 TBCs. *Surf. Coat. Technol.* **2019**, *371*, 262–275.

10. Lei, G.; Chenglong, Z.; Mingzhu, L.; Wei, S.; Zhaoyang, Z.; Fuxing, Y. Hot corrosion evaluation of $Gd_2O_3$-$Yb_2O_3$ co-doped $Y_2O_3$ stabilized $ZrO_2$ thermal barrier oxides exposed to $Na_2SO_4$ + $V_2O_5$ molten salt. *Ceram. Int.* **2017**, *43*, 2780–2785.

11. Cao, X.; Vassen, R.; Stoever, D. Ceramic Materials for Thermal Barrier Coatings. *J. Eur. Ceram. Soc.* **2004**, *24*, 1–10. [CrossRef]

12. Venkatesh, G.; Subramanian, R.; John Berchmans, L. Phase Analysis and Microstructural Investigations of Ce2Zr2O7 for High-Temperature Coatings on Ni-Base Superalloy Substrates. *High Temp. Mater. Proc.* **2019**, *38*, 773–782. [CrossRef]

13. Bahamirian, M.; Hadavi, S.M.M.; Farvizia, M.; Rahimipoura, M.R.; Keyvanic, A. Phase stability of ZrO2 9.5Y2O3 5.6Yb2O3 5.2Gd2O3 compound at 1100 °C and 1300 °C for advanced TBC applications. *Ceram. Int.* **2019**, *45*, 7344–7350. [CrossRef]

14. Miller, R.A. Thermal Barrier Coatings for Aircraft Engines: History and Directions. *J. Therm. Spray Technol.* **1997**, *6*, 35. [CrossRef]

15. Prakasam, M.; Valsan, S.; Lu, Y.; Balima, F.; Lu, W.; Piticescu, R.; Largeteau, A. Nanostructured Pure and Doped Zirconia: Syntheses and Sintering for SOFC and Optical Applications. *Sinter. Technol. Method Appl.* **2018**, 85. [CrossRef]

16. Ultramet Advanced Materials Solutions. Available online: https://ultramet.com/ceramic-protective-coatings/refractory-oxides/ (accessed on 29 May 2020).

17. Dong, T.S.; Wang, R.; Di, Y.L.; Wang, H.D.; Lia, G.L.; Fu, B.G. Mechanism of high temperature oxidation resistance improvement of double-layer thermal barrier coatings (TBCs) by La. *Ceram. Int.* **2019**, *45*, 9126–9135. [CrossRef]

18. Zhou, S.; Shen, S.; Fang, X.; Hou, Q.; Zhao, D. An accurate and rapid method to compare thermal insulation capacity of nine Gd-Yb-YSZ coatings. *Ceram. Int.* **2019**, *45*, 19910–19917. [CrossRef]

19. Boissonnet, G.; Chalk, C.; Nicholls, J.; Bonnet, G.; Pedraza, F. Thermal insulation of CMAS (CalciumMagnesium-Alumino-Silicates)- attacked plasma-sprayed thermal barrier coatings. *J. Eur. Ceram. Soc. Elsevier* **2020**, *40*, 2042–2049. [CrossRef]

20. Sasikumar, K.; Bharathikannan, R.; Raja, M.; Mohanbabu, B. Fabrication and characterization of rare earth (Ce. Gd, and Y) doped ZrO2 based metal-insulator-semiconductor (MIS) type Schottky barrier diodes. *Superlattices Microstruct.* **2020**, *139*, 106424. [CrossRef]

21. Mekala, R.; Deepa, B.; Rajendran, V. Preparation. characterization and antibacterial property of rare earth (Dy and Ce) doping on ZrO2 nanoparticles prepared by co-precipitation method. *Mater. Today Proc.* **2018**, *5*, 8837–8843. [CrossRef]

22. Deng, J.; Li, S.; Xiong, L.; Wang, J.; Yuan, S.; Chen, Y. Different thermal behavior of nanostructured $CeO_2$-$ZrO_2$ based oxides with varied Ce/Zr molar ratios. *Mater. Chem. Phys.* **2019**, *236*, 121767. [CrossRef]

23. Madhusudhanaa, H.C.; Shobhadevic, S.N.; Nagabhushanad, B.M.; Hari Krishnad, R.V.; Murugendrappa, M.; Nagabhushanaf, H. Structural Characterization and Dielectric studies of Gd doped $ZrO_2$ nano crystals Synthesized by Solution combustion method. *Mater. Today Proc.* **2018**, *5*, 21195–21204. [CrossRef]

24. Xiao, D.Q.; He, G.J.; Lv, G.P.; Wang, H.; Liu, M.; Gao, J.; Jin, P.; Jiang, S.S.; Li, W.D.; Sun, Z.Q. Interfacial modulation and electrical properties improvement of solution-processed $ZrO_2$ gate dielectrics upon Gd incorporation. *J. Alloy. Compd.* **2017**, *699*, 415–420. [CrossRef]

25. Guo, L.; Li, M.; Ye, F. Phase stabilityandthermalconductivityof $RE_2O_3$ ($RE\frac{1}{4}La$. Nd, Gd, Yb) and $Yb_2O_3$ co-doped $Y_2O_3$ stabilized $ZrO_2$ ceramics. *Ceram. Int.* **2016**, *42*, 7360–7365. [CrossRef]

26. Song, X.; Xie, M.; Mub, R.; Zhou, F.; Jia, G.; An, S. Influence of the partial substitution of $Y_2O_3$ with $Ln_2O_3$ (Ln = Nd, Sm, Gd) on the phase structure and thermophysical properties of $ZrO_2$–$Nb_2O_5$–$Y_2O_3$ ceramics. *Acta Mater.* **2011**, *59*, 3895–3902. [CrossRef]

27. Song, X.; Xie, M.; An, S.; Hao, X.; Mu, R. Structure and thermal properties of $ZrO_2$–$Ta_2O_5$–$Y_2O_3$–$Ln_2O_3$ (Ln = Nd. Sm or Gd) ceramics for thermal barrier coatings. *Scr. Mater.* **2010**, *62*, 879–882. [CrossRef]

28. Mikhailov, D.A.; Orlova, A.I.; Malanina, N.V.; Nokhrin, A.V.; Potanina, E.A.; Chuvil'deev, V.N.; Boldin, M.S.; Sakharov, N.V.; Belkin, O.A.; Kalenova, M.Y.; et al. A study of fine-grained ceramics based on complex oxides $ZrO_2$-$Ln_2O_3$ (Ln= Sm, Yb) obtained by Spark Plasma Sintering for inert matrix fuel. *Ceram. Int.* **2018**, *44*, 18595–18608. [CrossRef]

29. Wang, Y.; Chen, L.; Feng, J. Impact of $ZrO_2$ alloying on thermo-mechanical properties of $Gd_3NbO_7$. *Ceram. Int.* **2020**, *46*, 6174–6181. [CrossRef]

30. Chen, C.; Liang, T.; Guo, Y.; Chena, X.; Man, Q.; Zhang, X.; Zeng, J.; Ji, V. Effect of scandia content on the hot corrosion behavior of $Sc_2O_3$ and $Y_2O_3$ co-doped $ZrO_2$ in $Na_2SO_4$+$V_2O_5$ molten salts at 1000 °C. *Corros. Sci.* **2019**, *158*, 108094. [CrossRef]

31. Pitek, F.M.; Levi, C.G. Opportunities for TBCs in the $ZrO_2$–$YO_{1.5}$–$TaO_{2.5}$ system. *Surf. Coat. Technol.* **2007**, *201*, 6044–6050. [CrossRef]

32. Fan, W.; Baia, Y.; Wang, Z.Z.; Che, J.W.; Wang, Y.; Tao, W.Z.; Wang, R.J.; Liang, G.Y. Effect of point defects on the thermal conductivity of $Sc_2O_3$-$Y_2O_3$ costabilized tetragonal $ZrO_2$ ceramic materials. *J. Eur. Ceram. Soc.* **2019**, *39*, 2389–2396. [CrossRef]

33. Sun, L.; Guon, H.; Peng, H.; Gong, S.; Xu, H. Influence of partial substitution of $Sc_2O_3$ with $Gd_2O_3$ on thephase stability andthermalconductivityofSc2O3-doped $ZrO_2$. *Ceram. Int.* **2013**, *39*, 3447–3451. [CrossRef]

34. Fabrichnaya, O.; Wang, C.; Zinkevich, M.; Levi, C.G.; Aldinger, F.J. Phase equilibria and thermodynamic properties of the $ZrO_2$-$GdO_{1.5}$-$YO_{1.5}$ system. *J. Phase Equilib. Diff.* **2005**, *26*, 591–604. [CrossRef]

35. Fabrichnaya, O.; Savinykh, G.; Zienert, T.; Schreiber, G.; Seifert, H.J. Phase relations in the $ZrO_2$-$Sm_2O_3$-$Y_2O_3$-$Al_2O_3$ system: Experimental investigation and thermodynamic modelling. *Int. J. Mater. Res.* **2012**, *103*, 1469–1487. [CrossRef]

36. Andrievskaya, E.R.; Samelyuk, A.V.; Lopato, L.M. Reaction of cerium oxide with zirconium and yttrium oxides at 1250 °C. *Powder Metall. Met. Ceram.* **2002**, *41*, 63–71. [CrossRef]

37. Longo, V.; Podda, L. Relazioi tra le fasi allo stato solido nel sistema $CeO_2$–$ZrO_2$–$Y_2O_3$ tra 1700 e 1400 °C. *Ceramica (Florence)* **1984**, *37*, 18–20.

38. Li, L.; Van der Biest, O.; Wang, P.L.; Vleugels, J.; Chen, W.W. Estimation of the phase diagram for the ZrO2-Y2O3-CeO2 system. *J. Eur. Ceram. Soc.* **2001**, *21*, 2903–2910. [CrossRef]

39. Fabrichnaya, O.B.; Savinykh, G.; Schreiber, G.; Dopita, M.; Seifert, H.J. Experimental investigation and thermodynamic modelling in the $ZrO_2$–$La_2O_3$–$Y_2O_3$ system. *J. Alloys Compd.* **2010**, *493*, 263–271. [CrossRef]

40. Fabrichnaya, O.B.; Savinykh, G.; Schreiber, G.; Seifert, H.F. Phase relations in the $ZrO_2$ –$Nd_2O_3$ –$Y_2O_3$ system: Experimental study and CALPHAD assessment. *Int. J. Mater. Res.* **2010**, *101*, 1354–1360. [CrossRef]

41. Zhang, D.; Gong, S.; Huibin, X. Thermal cycling behaviors of thermal barrier coatings on intermetallic Ni3Al based superalloy. *Surf. Coat. Technol.* **2003**, *168*, 78–83. [CrossRef]

42. Boissonnet, G.; Chalk, C.; Nicholls, J.R.; Bonnet, G.; Pedraza, F. Phase stability and thermal insulation of YSZ and erbia-yttria co-doped zirconia EB–PVD thermal barrier coating systems. *Surf. Coat. Technol.* **2020**, *389*, 125566. [CrossRef]

43. Cao, X.; Vassen, R.; Jungen, W.; Schwartz, S.; Tietz, F.; Stöver, D. Thermal stability of lanthanum zirconate plasma-sprayed coating. *J. Am. Ceram. Soc.* **2001**, *84*, 2086–2090. [CrossRef]

44. Vassen, R.; Cao, X.Q.; Tietz, F.; Basu, D.; Stöver, D. Zirconates as new materials for thermal barrier coatings. *J. Am. Ceram. Soc.* **2000**, *83*, 2023–2028. [CrossRef]

45. Mercer, C.; Williams, J.R.; Clarke, D.R.; Evans, A.G. On a ferroelastic mechanism governing the toughness of metastable tetragonal-prime (t′) yttria-stabilized zirconia. *Proc. R. Soc. A* **2007**, *463*, 1393–1408. [CrossRef]

46. Guo, X.Y.; Li, L.; Park, H.M.; Knapp, J.; Jung, Y.G.; Zhang, J. Mechanical properties of layered $La_2Zr_2O_7$ thermal barrier coatings. *J. Therm. Spray Technol.* **2018**, *27*, 1–10. [CrossRef]

47. Zhang, D.; Liao, K.; Yu, Y.; Tian, Z.; Cao, Y. Microstructure and thermal & mechanical properties of La2Zr2O7@YSZ composite ceramic. *Ceram. Int.* **2020**, *46*, 4737–4747.

48. Karaoglanli, A.C.; Doleker, K.M.; Ozgurlu, Y. Interface failure behavior of yttria stabilized zirconia (YSZ), $La_2Zr_2O_7$, $Gd_2Zr_2O_7$, YSZ/$La_2Zr_2O_7$ and YSZ/$Gd_2Zr_2O_7$ thermal barrier coatings (TBCs) in thermal cyclic exposure. *Mater. Charact.* **2020**, *159*, 110072. [CrossRef]

49. Doleker, K.M.; Ozgurluk, Y.; Karaoglanl, A.C. Isothermal oxidation and thermal cyclic behaviors of YSZ and doublelayered YSZ/La2Zr2O7 thermal barrier coatings (TBCs). *Surf. Coat. Technol.* **2018**, *351*, 78–88. [CrossRef]

50. Naga, S.M.; Awaad, M.; El-Maghraby, H.F.; Hassan, A.M.; Elhoriny, M.; Killinger, A.; Gadow, R. Effect of La2Zr2O7 coat on the hot corrosion of multi-layer thermal barrier coatings. *Mater. Des.* **2016**, *102*, 1–7. [CrossRef]

51. Wang, C.; Wang, Y.; Fan, S.; You, Y.; Wang, L.; Yang, C. Optimized functionally graded $La_2Zr_2O$/8YSZ thermal barrier coatings. *J. Alloy. Compd.* **2015**, *649*, 1182–1190. [CrossRef]

52. Ozgurluk, Y.; Doleker, K.M.; Karaoglanli, A.C. Hot corrosion behavior of YSZ, $Gd_2Zr_2O_7$ and YSZ/$Gd_2Zr_2O_7$ thermal barrier coatings exposed to molten sulfate and vanadate salt. *Appl. Surf. Sci.* **2018**, *438*, 96–113. [CrossRef]

53. Amaya, C.; Aperador, W.; Caicedo, J.C.; Espinoza-Beltrán, F.J.; Muñoz-Saldaña, J.; Zambrano, G.; Prieto, P. Corrosion study of Alumina/Yttria-Stabilized Zirconia ($Al_2O_3$/YSZ) nanostructured Thermal Barrier Coatings (TBC) exposed to high temperature treatment. *Corros. Sci.* **2009**, *51*, 2994. [CrossRef]

54. Rösemann, N.; Ortner, K.; Petersen, J.; Schadow, T.; Bäker, M.; Bräuer, G.; Rösler, J. Influence of bias voltage and oxygen flow rate on morphology and crystallographic properties of gas flow sputtered zirconia coatings. *Surf. Coat. Technol.* **2015**, *276*, 668.

55. Rösemann, N.; Ortner, K.; Bäker, M.; Petersen, J.; Bräuer, G.; Rösler, J. Influence of the Oxygen Flow Rate on Gas Flow Sputtered Thermal Barrier Coatings. *J. Ceram. Sci. Technol.* **2018**, *9*, 29.

56. Noor-A-Alam, M.; Choudhuri, A.R.; Ramana, C.V. Effect of composition on the growth and microstructure of hafnia–zirconia based coatings. *Surf. Coat. Technol.* **2011**, *206*, 1628. [CrossRef]

57. Portinha, A.; Teixeira, V.; Carneiro, J.; Costa, M.F.; Barradas, N.P.; Sequeira, A.D. Stabilization of $ZrO_2$ PVD coatings with $Gd_2O_3$. *Surf. Coat. Technol.* **2004**, *107*, 188–189. [CrossRef]

58. Andritschky, M.; Teixeira, V.; Rebouta, L.; Buchkremer, H.P.; Stover, D. Adherence of combined physically vapor-deposited and plasma-sprayed ceramic coatings. *Surf. Coat. Technol.* **1995**, *101*, 76–77.

59. Yao, J.; He, Y.; Wang, D.; Peng, H.; Guo, H.; Gong, S. Thermal barrier coatings with $(Al_2O_3–Y_2O_3)/(Pt\ or\ Pt–Au)$ compositebond coat and 8YSZ top coat on Ni-based superalloy. *Appl. Surf. Sci.* **2013**, *286*, 298. [CrossRef]

60. Budinovskiia, S.A.; Matveeva, P.V.; Smirnova, A.A.; Chubarov, D.A. Thermal-Barrier Coatings with an External Magnetron-Sputtered Ceramic Layer for High-Temperature Nickel Alloy Turbine Blades. *Russ. Metall. (Met.)* **2019**, *6*, 617–623. [CrossRef]

61. Tabatabaeian, M.R.; Rahmanifard, R.; Jalili, Y.S. The study of phase stability and thermal shock resistance of a Scandia–Ceria stabilized zirconia as a new TBC material. *Surf. Coat. Technol.* **2019**, *374*, 752–762. [CrossRef]

62. Byrappa, K.; Yoshimura, M. *Handbook of Hydrothermal Technology*, 2nd ed.; Elsevier: Amsterdam, The Netherlands, 2013.

63. Piticescu, R.R.; Corban, M.; Grilli, M.L.; Balima, F.; Prakasam, M. Design of new coatings and sintered materials based on mixed rare earth oxides. *J. Nucl. Res. Dev.* **2019**, *18*, 18–23.

64. Piticescu, R.R.; Malic, B.; Kosec, M.; Motoc, A.; Monty, C.; Soare, I.; Kosmac, T.; Daskobler, A. Synthesis and sintering behaviour of hydrothermally synthesized YTZP nanopowders for ion-conduction applications. *J. Eur. Ceram. Soc.* **2004**, *24*, 1941–1944. [CrossRef]

65. Ionescu, G.; Manoliu, V.; Alexandrescu, E.; Stefan, A.; Mihailescu, A. The behavior of multilayer ceramic protecting at thermal shock. *Incas Bull.* **2013**, *5*, 25–32.

 *metals*

*Review*

# A Bibliometric Analysis of the Publications on In Doped ZnO to be a Guide for Future Studies

**Mehmet Yilmaz** [1,2,*], **Maria Luisa Grilli** [3] **and Guven Turgut** [4,5]

1   Department of Science Teaching, K. K. Education Faculty, Atatürk University, 25240 Erzurum, Turkey
2   Advanced Materials Research Laboratory, Department of Nano science and Nano engineering, Graduate School of Natural and Applied Sciences, Ataturk University, 25240 Erzurum, Turkey
3   ENEA-Italian National Agency for New Technologies, Energy and Sustainable Economic Development, Energy Technologies Department, Casaccia Research Centre, Via Anguillarese 301, 00123 Roma, Italy; marialuisa.grilli@enea.it
4   Department of Basic Sciences, Science Faculty, Erzurum Technical University, 25240 Erzurum, Turkey; guventrgt@gmail.com
5   Photonic Group, High Technology Research Centre (YUTAM), Erzurum Technical University, 25240 Erzurum, Turkey
*   Correspondence: mehmetyilmaz@atauni.edu.tr or yilmazmehmet32@gmail.com; Tel.: +90-442-231-4155

Received: 7 April 2020; Accepted: 29 April 2020; Published: 1 May 2020

**Abstract:** This study aims to examine the studies regarding In doped ZnO published in the Web of Science database. A total of 777 articles were reached (31 March 2020). The articles were downloaded for the bibliometric analysis and collected in a file. The file was uploaded to VOSViewer programme in order to reveal the most used keywords, words in the abstracts, citation analyses, co-citation and co-authorship and countries analyses of the articles. The results showed that the most used keywords were "ZnO", "photoluminescence", "optical properties", "thin films" and "doping". These results indicate that the articles mostly focus on some characteristics of In doped ZnO thin films such as structural, optical and electrical features. When the distribution of the number of articles using the keywords by year was searched, it was found that recent articles focus mainly on synthesis of In doped ZnO film via chemical routes such as sol-gel and hydrothermal syntheses, and on ZnO-based device applications such as solar cells and gas sensors. The most used keywords were also found to be films, X-ray, glass substrate, X-ray Diffraction (XRD), spectra and layer. These results indicate that the studies mostly focus on In doped ZnO thin films as transparent conductive oxide (TCO) material used in device applications like solar cells. In this context, it was found that structural, topographical, optical, electrical and magnetic properties of In doped ZnO films were characterized in terms of defected structure or defect type, substrate temperature, film thickness and In doping content. When the distribution of these words is shown on a year-by-year basis, it is evident that more recent articles tend to focus both on efficiency and performance of In doped ZnO films as TCO in solar cells, diodes and photoluminescence applications both on nanostructures, such as nanoparticles, and nanorods for gas sensor applications. The results also indicated that Maldonado and Asomoza were the most cited authors in this field. In addition, Major, Minami and Ozgur were the most cited (co-citation) authors in this field. The most cited journals were found to be Thin Solid Films, Journal of Materials Science Materials in Electronics and Journal of Applied Physics and, more recently, Energy, Ceramics International, Applied Physics-A, Optik, Material Research Express, ACS Applied Materials and Interfaces and Optical Materials. The most co-cited journals were Applied Physics Letters, Thin Solid Films, Journal of Applied Physics, Physical Review B, and Applied Surface Science. Lastly, the countries with the highest number of documents were China, India, South Korea, USA and Japan. Consequently, it is suggested that future research needs to focus more on synthesis and characterization with different growth techniques which make In doped ZnO suitable for device applications, such as solar cells and diodes. In this context, this study may provide valuable information to researchers for future studies on the topic.

**Keywords:** In doped ZnO; bibliometric analysis; thin films; metal oxides

---

## 1. Introduction

Indium, whose abundance in earth is similar to silver, is one of the rare metals in the earth's crust and is nearly found as a trace element in other minerals. The importance of indium metal and its salts in organic synthesis has been recognized due to the unusual carbon–carbon bond promotion discovered in recent years, its rearrangement reactions and its performance in various beneficial reactions [1,2]. Indium is a metal with a silver-white color and a low melting point, belonging to the metal group, and it does not have the electronic structure of inert gases when it loses its outermost orbital electrons. Therefore, indium is not reactive like typical metals. Compared with lead, it is softer and malleable, but does not oxidize depending on the temperature [3,4]. Indium has also a very special functional application because it is frequently used in semiconductor based devices, thermistors and optical devices. This not only increases the popularity of indium, but also increases the need for indium in these applications. In other words, due to the demand of companies operating in the solar and wind sectors, as Hoffman et al. [5] mentioned in their studies, the demand for indium in addition to tellurium, gallium, dysprosium and neodymium is expected to increase significantly in the coming years. Considering the increasing energy need with the development of technology, it can be concluded that an element that has the status of rare elements in the earth's crust, like indium, should be used carefully. Moreover, indium has been listed recently as a critical raw material (CRM) by the European Union, due to its high supply risk and its high economic importance [6]. Indium is mainly used as indium tin oxide (ITO), which is the transparent conductive oxide of choice in a wide range of applications, from solar cells to LED panels. ITO's main constituent is indium (ITO contains approximately 78% indium), and therefore, researchers have focused recently on searching for ITO-alternative materials for optoelectronic applications. Among the proposed alternatives, zinc oxide steps forwards due to its unique properties such as its wide and direct band gap ($\approx$3.3 eV), large exciton binding energy (60 meV), highly transparency in visible range and low cost [7–9]. Additionally, easy tunability of its morphology for the application type makes it a very suitable material for solar cells, energy hydrogen conversion devices and sensors [10,11]. From studies on ZnO films, it can be seen that pure ZnO thin films are not very stable during chemisorption and desorption of oxygen due to variable surface conductance [12]. An efficient way to reduce this disadvantage is doping [13]. The properties of ZnO can be controlled by doping and thus tailored for the desired applications. For example, although ZnO has natural n-type electrical conductivity associated with zinc interstitial and oxygen deficiency, ZnO with p-type electrical conductivity can be obtained by doping with La and As [14]. In reality, the growth of stable ZnO films with p-type conductivity is very difficult. Therefore, it is important to determine suitable dopant materials before the experimental procedure. Feng and Xia [15] found that although a large amount of dopants such as N, P, Sb, Co may be used to obtain ZnO with p-type doping, it is difficult to obtain ZnO with stable p-type conductivity due to their low solubility and deep level of acceptor characteristics. However, with the addition of arsenic to ZnO, a complex acceptor (AsZn-2Vzn) with relatively high ionization energy (137 meV) is formed. Thus, ZnO with stable p-type conductivity is obtained. Al, B, In, Ga are generally used to obtain n-type degenerate electrical semiconductors [16], due to the increase of free electrons concentration. Therefore, ZnO may be doped with several elements to achieve both n- and p-type conductivity. In order to reveal the general research tendency in the study of ZnO, a bibliometric mapping analysis was performed. Bibliometrics is an emerging cross-disciplinary analysis based on statistic and mathematic tools to map the state of the art and the development in a given area of scientific knowledge [17–20]. It allows us to identify essential information on a particular topic based on the analysis of citations, co-citations, geographical distribution and word frequency, etc., revealing the current research situation and giving insight into the future trends. In our study, the bibliometrics analysis (5439 articles were reached with keywords of ("dop*") AND (ZnO OR "zinc oxide")) was obtained by analyzing the articles published in the last three years on doped ZnO and examining them in terms of the distribution of the most frequently mentioned words in the abstracts. The results are reported in Figure 1.

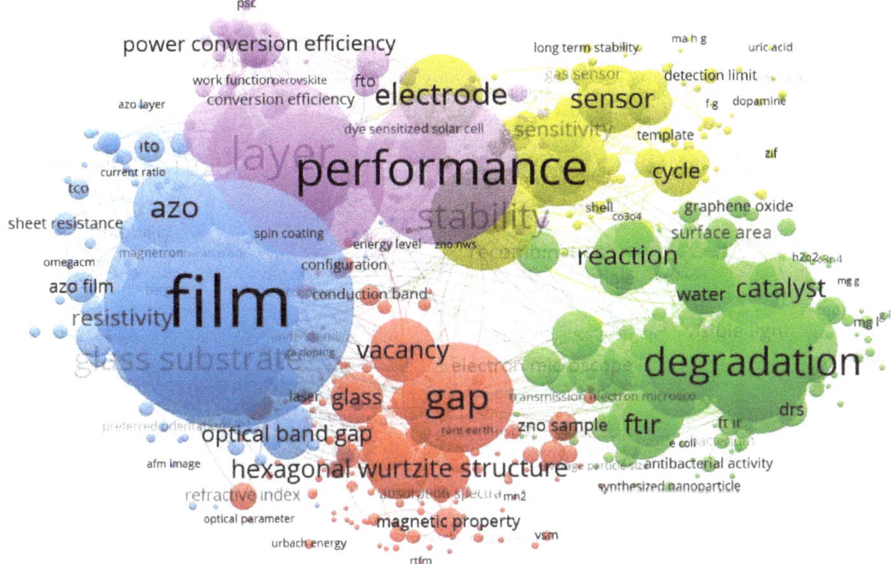


**Figure 1.** Most used words in abstract sections about doped ZnO articles.

From Figure 1, it can be observed that the frequency of the words "film", "performance", "device", "electrode", "power conversion efficiency", "sensor", "degradation" is high. This means that ZnO has a very wide range of applications. Nevertheless, the bibliometric analysis evidenced high interest towards Al doped ZnO (AZO), which is the most investigated ITO alternative, and In was selected in this review as the ZnO dopant, because it has the advantage of less reactivity and higher oxidation resistance than other additives of the same group [21]. Indium is, at present, listed among CRMs, however its content in indium doped zinc oxide (IZO) is quite low with respect to ITO. It is well known that the characteristics of ZnO depend both on the dopant content as well as on the growth condition and the growth technique. In some studies, [22,23] the dopant was limited in the range of 4 at.%–5 at.% because at a further doping ratio, the deterioration of the ZnO crystal structure or higher resistivity were found. Moreover, the electronegativity of foreign ions and standard electrode potentials (SEP) also affect carrier suppression abilities [24]. In this context, considering the electronegativity (1.78), standard electrode potential (−0.25 V) and ionic radius of indium (0.08 nm for $In^{3+}$), In can be considered as a suitable dopant for ZnO in applications requiring high conductivity and optical transmission [25,26]. Therefore, in applications which require the use of indium doped tin oxide, the use of indium doped ZnO films with same optoelectronic properties can strongly reduce the use of indium. Shaheera et al. [27] grew In doped ZnO via the radio frequency (RF) sputtering technique and showed that the samples analyzed in their study can be used in white light source and LED device applications. In another study, Yu et al. [28] obtained oxygen defect rich In doped ZnO nanostructures by using a facile solution method followed by an annealing process, and proved that the photocatalytic features of ZnO improved as a result of $In^{3+}$ addition into the ZnO. Bhatia and Verma [29] also investigated the performance of In doped ZnO films obtained by the rapid sol-gel dip coating technique in gas sensing application and showed that In doped ZnO films are good candidates for $NO_2$ gas sensing due to their quick response time of 65 s and a fast recovery time of 87 s. Kyaw et al. [30] used sol-gel coated indium doped ZnO films and investigated film performance in organic solar cell applications. They obtained films with $5.54 \times 10^{-1}$ $\Omega cm^{-1}$ resistivity and 80% optical transparency which exhibited good performance as buffer layers in organic solar cell applications. In summary, as can be concluded from the aforementioned discussion, In doped ZnO has been widely used for different applications. Therefore, it is important to examine literature in detail for future studies. In our work, the studies regarding indium doped ZnO published to date in the Web of Science database were analyzed by the bibliometric analysis method. Some insights into the characteristics of IZO films are given and discussed on the basis of data from literature. Therefore, since the current study covers important studies about this subject and includes basic discussions, it will help researchers by providing a detailed background to guide their work.

## 2. Materials and Methods

This study aims to examine the articles regarding In doped ZnO published in Web of Science database. The keywords "indium dope*" OR "in dope*" OR "influence of in dope*" OR "influence of indium dope*" OR "effect* of in dope*" OR "effect* of indium dope*" OR "addition of indium dope*" OR "addition of In dope*" AND "ZnO*" OR "zinc oxide" were entered to the "topic" sections and 900 articles were reached (31 March 2020). The language was selected as "English" and document type was fixed as "article". Then, 777 articles were obtained for analysis. These articles were downloaded as tab-delimited (Win format) for the bibliometric analysis (full record and cited references format). The file was uploaded to VOSViewer programme in order to reveal the most used keywords, words in the abstracts, citation analyses, co-citation and co-authorship in countries analyses in the articles. The process of article selection is summarized in Figure 2.

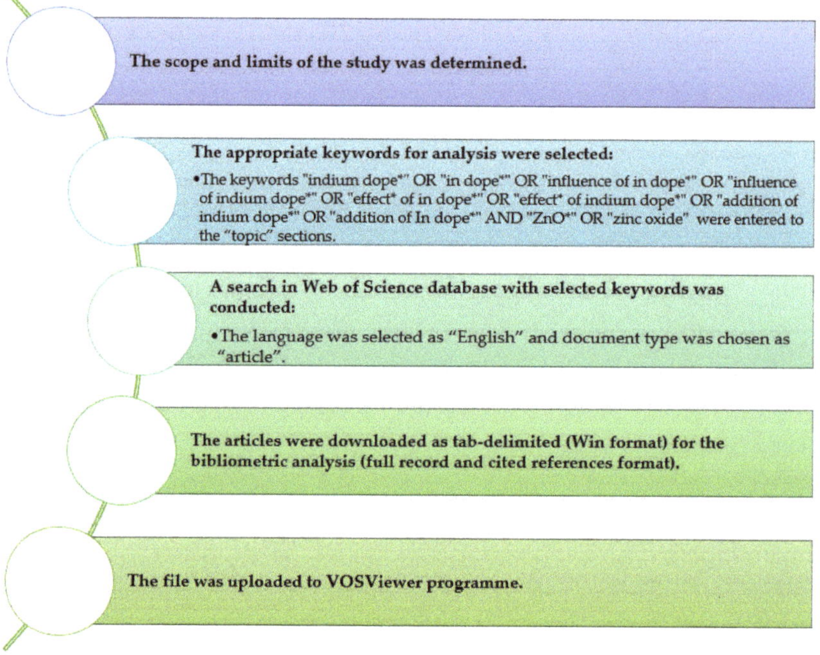



**Figure 2.** The process of article selection.

## 3. Results

In order to create a map based on text data for the most used keywords, co-occurrence analysis was used and keywords were chosen by the selected authors. The minimum number of occurrences of a keyword was set as five, and the number of keywords to be selected was automatically obtained as 70. The map created is illustrated in Figure 3.

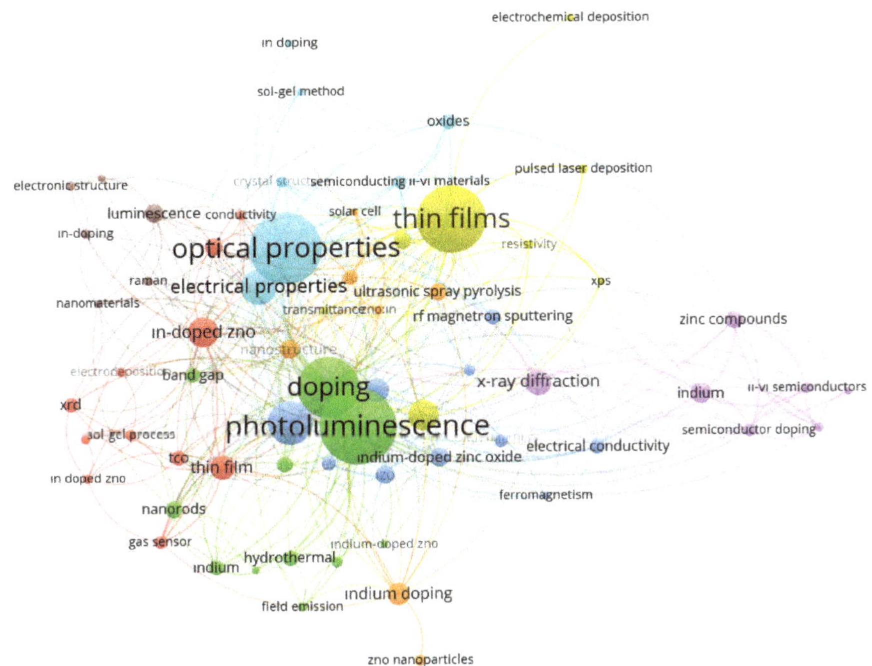


**Figure 3.** The most used keywords in the articles related to In doped ZnO.

"ZnO" and "zinc oxide" are the most used keywords, as expected. These keywords were removed from the map to highlight the other most used keywords, which are: 'photoluminescence' (f = 47), "optical properties" (f = 44), "thin films" (f = 42) and "doping" (f = 38). Results indicate that the publications mostly focus on some characteristics of In doped ZnO thin films such as structural, optical and electrical features. Some of the studies related to In doped ZnO films grown by different techniques were analyzed, and IZO main features were reported in Table 1.

**Table 1.** Comparison of In doped ZnO films grown by different techniques in terms of structural, optical and electrical characteristics.

| The Study (As a Function of Increasing "In" Rate) | Growth Technique | Crystallite Size (nm) | Optical Band Gap (eV) | Resistivity ($\Omega$cm) |
|---|---|---|---|---|
| Filali et al. [31] | Ultrasonic spray pyrolysis | 26–28; 21–23 | 3.31; 3.27 | - |
| Khalfallah et al. [32] | RF magnetron sputtering | 24.91; 31.06 | 3.27; 3.36 | $5.35 \times 10^{-3}$ |
| Thambidurai et al. [33] | Sol-gel | 22.4; 7.3 | 3.12; 3.23 | $7.98 \times 10^{-3}; 2.40 \times 10^{-3}$ |
| Benhaliliba et al. [34] | Ultrasonic spray pyrolysis technique | 39; 24 | 3.18; 3.21 | ~385; ~8.35 |
| Benouis et al. [35] | Spray pyrolysis | 26; 55 | 3.28; 3.35 | $17 \times 10^{-3}; 6 \times 10^{-3}$ |
| Lee et al. [36] | Pyrosol method | ~49; ~62 | – | $1.3 \times 10^{-2}; 3.0 \times 10^{-3}$ |
| Silva-Lopez et al. [21] | RF magnetron sputtering | 16.4; 16.0 | $3.2 \pm 0.1$ | $2.9 \pm 0.7 \times 10^{-3}$ |
| Rambu et al. [37] | Thermal evaporation | 29; 24 | 3.32; 3.38 | $4.12 \times 10^{-3}; 4.26$ |
| Tang et al. [38] | MOCVD | 38.2; 20.1 | ~3.25; ~3.18 | ~30; ~6 $\times 10^{-3}$ |
| Jongthammanurak et al. [39] | Ultrasonic spray pyrolysis technique | 18.7; 25.4 | 3.300; 3.293 | 0.521; 0.176 |

According to the studies reported in Table 1, the characteristics of ZnO depend on both growth technique and In doping content. In other words, ZnO, which is characterized by the hexagonal wurtzite structure, may exhibit different characteristics due to defects such as zinc interstitials and oxygen vacancies, which depend on the growth technique and growth parameters. Similar results were found also by many other authors [40,41]. Therefore, it can be suggested that the evaluation of ZnO's characteristics should be made taking the growth technique into consideration. For example, in the studies [31,34], which investigate the variations in the characteristics of ZnO grown onto glass substrates by the ultrasonic spray pyrolysis technique as a function of In content, the authors found that samples generally have (002) crystallographic direction along c-axis. This preferential crystallization can be explained by the fact that the atoms have enough energy to spread to this region since the direction of the hexagonal ZnO structure (002) requires the lowest surface energy, so it can be thought that (002) orientation is the most optimal growth orientation for ZnO films [42–44]. On the other hand, for films grown by the spray pyrolysis technique, Prasada Rao and Santhoshkumar [45] found a highly oriented (100) ZnO crystallization. They explained this preferred orientation in terms of the amount of ethanol percentage, which affects the films' crystallinity, the orientation of the crystallites in the films and the films' morphology. The preferred orientation is explained in terms of nucleation and final growth orientation, both of which result from the nucleation at the film/substrate interface [46]. Similar interpretations were made by other researchers [47]. Other examples of the preferred orientation of ZnO depending on solution chemistry or growth technique are reported below. Dimitrov et al. [48] showed that if the solution is prepared by adding $CH_3COOH$ to the zinc ammonia precursor, ZnO with the (002) orientation can be obtained, while ZnO with the (100) orientation can be obtained if the zinc ammonia precursor is prepared with $HNO_3$ or HCl. Additionally, it is possible to obtain ZnO films with a (101) preferred orientation by using RF magnetron sputtering. In summary, it can be said that many factors, such as substrate, heat treatment condition, solution chemistry, directly affect the crystallinity and thus the preferred orientation of the samples [49]. The most probable causes of the three possible orientations for ZnO are discussed above and presented to researchers as a preliminary information for future studies. However, it should be noted that the variation in the preferred orientation affects the properties of the samples. Therefore, it is very important to prepare ZnO to fit desirable purposes. For example, the dark current of ZnO obtained by the solution-based spray pyrolysis technique with a preferred orientation (100) displays a dark current smaller than (002) one [48]. In reference [48], authors found that ZnO films with orientation (100) reach the maximum dark current five seconds earlier than films with orientation (002), showing improved switching performances. Therefore, for researchers who want to examine the sensitivity of ZnO to ammonia, ethanol, acetone and water vapors, it may be suggested to grow ZnO films with preferred (100) orientation, which have much lower conductivity and are sensitive to vapor-based changes. After discussing the changes in preferred orientation, it is also necessary to mention the reasons for possible changes in peak intensity due to external doping. Peak intensity varies depending on indium doping. Fluctuations in the (002) peak appear as the result of deterioration of the crystal structure of ZnO due to In content. Defects induced by doping affect the structural, morphological, optical, electrical and magnetic properties of ZnO [50,51]. From the studies reported in Table 1, it can be observed that, in some cases, crystallite size increases with increasing In content, while in others an opposite tendency is observed. The differences between them can be summarized in terms of following possible reasons:

- Factors, such as increase in the density of ZnO grains, the presence of elastic stress, the presence of In atoms as an interstitial atom in the ZnO crystal structure, lead to a decrease in the crystalline size of ZnO [31,37,38].
- Generally, increasing crystallite size is linked with the ionic radii of the dopant and host ions. That is, increased crystallite sizes are likely to occur frequently, especially when larger atoms are used as the dopant. In this case, larger grains will be expected because hosted atoms with larger size replace the Zn atoms in the lattice. Additionally, some researchers mentioned that the

morphology of the samples varies because of particle agglomeration due to indium content and therefore bigger particles were obtained [32,35,36,39,52].

It is known that In content and In dependent structural and morphological characteristics directly affect ZnO's optoelectronic properties. As stated in the study of Hamberg and Granqvist [53], band-gap shift in a semiconductor is defined in terms of two competing mechanisms. The Burstein–Moss effect is generally used to define band-gap widening in heavily doped semiconductors, while electron–electron or electron–ion scattering is used for explaining band-gap narrowing [54,55]. The Burstein–Moss shift effect can be described as follow [56]:

$$\Delta E_g^{BM} = \frac{\hbar^2}{2m^*}\left(3\pi^2 n\right)^2 \tag{1}$$

where n and $m^*$ are the carrier concentration and effective mass, respectively. From this Equation, it is possible to infer theoretically the Burstein-Moss shift, once known the carrier concentration. However, when the ZnO matrix is occupied by an ion belonging to the IIIB family, the electron concentration in ZnO increases. Similar to the study made by Cao et al. [55], the generation of free electrons in ZnO as a result of In doping can be expressed by using the following equations:

$$ZnO \leftrightarrow ZnO_{1-x} + \frac{x}{2}\,O_2(g) \tag{2}$$

$$O_O \leftrightarrow \frac{1}{2}O_2(g) + V_O \tag{3}$$

In Equation (2), $ZnO_{1-x}$ x term takes into account the neutral oxygen vacancies $V_O$ formed during the growth process of ZnO based on Equation (3). These neutrally formed oxygen vacancies may be singly ionized ($V_O^{\bullet}$) or doubly ionized ($V_O^{\bullet\bullet}$), and are called paramagnetic and diamagnetic species, respectively. The occurrence of an electron as a result of single or doubly ionizing process may be described as follows [57]:

$$V_O \leftrightarrow V_O^{\bullet} + e' \tag{4}$$

$$V_O \leftrightarrow V_O^{\bullet\bullet} + 2e' \tag{5}$$

where $e'$ is the electron in the conduction band of ZnO. From Equations (2–5), an interpretation on how to increase electron concentration in In doped ZnO can be made. Each Zn atom contributes 2/3 $e'$ to the adjacent O or O vacancy to achieve ZnO with n-type electrical conductivity. In this condition, if the Zn atom is substituted by In in the ZnO matrix, every In atom gives 3/3 $e'$ to the close O or O vacancy. Therefore, electron concentration increases because of In impurities in ZnO. From the aforementioned discussion, it can be concluded that if ZnO is doped with elements such as "In, Ga, Al Sn" belonging to IIIB and IVB of the periodic table, these elements act as donors, while IA group elements like Li act as acceptors. Therefore, the Burstein–Moss effect can be seen an effective way to explain the band-gap widening effect in ZnO as a result of the presence of In impurities. As mentioned before, the In addition-based defects level was mostly predicted using photoluminescence (PL) measurements. Generally, two emission peaks were observed in the PL spectra of ZnO. A strong peak is located at ~383 nm, while the weak one is located at ~550 nm. The broad peak located in the range of 500–600 nm can be associated with the non-stoichiometric intrinsic defects like zinc vacancies [58]. Moreover, there is a significant change in PL spectra of ZnO because of In doping. According to S.Y. Lim et al.'s [59] study, it can be observed that the intensity of UV emission is increased, while the peak shifts to higher wavelength as a result of increasing In content. This variation in visible luminescence can be related to defect induced emission resulting from increasing defect density. The broadening of the UV peak in PL spectra as a result of the indium addition can also be attributed to the band tail effect, which can be induced by indium impurities into the ZnO lattice.

It is well known that, in a transparent conductive oxide (TCO), the optical properties are strongly correlated to the electrical conductivity and both should be optimized for optoelectronic applications.

Therefore, it is necessary to investigate the changes in electrical properties of ZnO films in terms of In content. This case is also seen in the cloud displayed in Figure 5. Additionally, in Figure 5, frequently passing keywords that indicate electrical properties of indium doped ZnO films, such as electrical property, device, carrier concentration, electrical resistivity and low resistivity, also indicate the importance of indium doped zinc oxide films in device applications. Most studies indicate that In doped ZnO films exhibit n-type electrical conductivity [60–62]. Additionally, the electrical conductivity of ZnO varies depending on the In content. For example, the sheet resistance of ZnO thin films obtained by Kumar et al. [63] by the chemical spray pyrolysis method was measured as 74 ohms/square and they associated this low resistance with the high amount of zinc metal found in ZnO (Zn/O ratio was 1.77). The electrical resistivity of ZnO films generally decreases with indium incorporation due to the increase in free electron concentration. That is, a large number of indium atoms can ionize in the form of $In^{3+}$ and replace $Zn^{2+}$ in the ZnO crystal structure, thereby contributing one free electron from each indium atom. The grain boundary scattering of electrons causes an increase in resistivity [22,61]. However, at high In contents, In segregation may occur and an increase in IZO resistance is observed [64,65]. This case is generally due to the fact that not all indium atoms contribute to conductivity with a free electron [66]. Additionally, similarly to Peng et al.'s study [67], increasing resistivity in higher indium doping content can be linked with the segregation of indium atoms in non-crystalline regions on the grain boundaries.

Table 2 compares the resistivity of In doped ZnO films obtained by different researchers with different techniques. Based on this, the lowest resistivity value for IZO films were obtained with the dc-reactive sputtering technique, while the highest resistivity value was obtained with the spray pyrolysis technique. Additionally, the resistivity value is related to the carrier concentration. In this sense, starting from the studies given in Table 2, it can be concluded that with the increase of In content (considering the doping rates to be determined depending on the technique used), the concentration of the electrical carrier increases and therefore the resistivity decreases.

**Table 2.** Comparison of resistivity of In doped ZnO films obtained by different techniques.

| Sample Name | Growing Tech. | Doping Ratio (at%) | Resistivity ($\Omega$cm) |
|---|---|---|---|
| IZO 3 [68] | Spray pyrolysis | 3 | $3.48 \times 10^{-2}$ |
| IZO33 [69] | Dip coating | 0.33 | 1.48 |
| $(Zn_{1-x}In_x)O$ [70] | Solid-state reaction | 0.02 | $1.884 \times 10^{-3}$ |
| IZO-6 [71] | Ultrasonic spray pyrolysis | 4 | $5.7 \times 10^{-3}$ |
| In:ZnO [72] | Spray Pyrolysis | 2.5 | $1.77 \times 10^{-2}$ |
| ZnO:In [73] | PLD | - | $10^{-1}$ |
| IZO [54] | Dc reactive co-sputtering | 3.2 | $3.6 \times 10^{-4}$ |
| In:ZnO [74] | Hydrothermal | 0.55 | $1.5 \times 10^{-2}$ |
| ZnO-1 [75] | Modified S-gun magnetron sputtering | 6 | $1.08 \times 10^{-3}$ |
| ZnO:In [36] | Pyrosol spray method | 3 | $3 \times 10^{-3}$ |

Additionally, some studies report on resistivity increases at high In content. For example, Benouis et al. [35] found that IZO resistance tends to increase when In is higher than 2 at%, and that possible structural disorders occur as a result of doping. The authors explained the degradation of the electrical resistivity as due to the accumulation of In in indium hydroxide $In(OH)_3$ forming at the grain boundaries, and to the fact that chlorine functions as a trap for electrons [76]. In addition, similarly to what occurs in other TCOs, by increasing film thickness, the concentration of charge carriers increases, thus increasing the conductivity of the material and reducing its transparency. Therefore, an optimal compromise must be acquired between optical transparency and electrical conductivity, with the optical and electrical characteristics correlated with an inverse relationship. It would be preferable to increase the thickness rather than increasing the doping level of a TCO, in order to decrease the sheet resistance of the film while maintaining free carrier absorption as low as possible.

The effect of post deposition annealing on the optical and electrical properties of IZO films has been investigated by several authors. For instance, Guo-Ping et al. [77] found that deposited IZO films grown by RF sputtering have carrier concentrations in the range of $10^{19}$ cm$^{-3}$, which increased up to

$4.37 \times 10^{20}$ cm$^{-3}$ with a 30 min annealing treatment at 450 °C in argon atmosphere. As a result of this, they obtained IZO films with $2.5 \times 10^{-3}$ Ωcm resistivity. Additionally, Barquinha et al. [78] found a decrease of the electrical resistivity of IZO films upon annealing. However, there are many other works reporting on the increase of IZO films resistance as a result of annealing [70]. Moreover, considering the studies reported in Tables 1 and 2, it may be observed that samples grown with the same method by different authors have different characteristics. Samples prepared by chemical synthesis may have different precursor solutions and different dopant concentrations, which lead to different electrical and optical properties. For films deposited by Physical Vapour Deposition techniques, several growth parameters (sputtering pressure, oxygen partial pressure, substrate temperature, target to substrate distance, RF or DC powers, magnetron, etc.) affect films' performances. From the analyzed works, we can summarize that best film performances are achieved generally by physical vapor deposition (PVD) techniques. In addition, differences may also occur in ZnO-based devices. For example, diodes fabricated in the same way may have different diode characteristics, due to different thicknesses of the oxide layer in the metal semiconductor interface and/or due to the non-homogeneous distribution of ZnO as the interface layer [79]. The above-mentioned discussion shows the variety of characteristics of indium doped zinc oxide films, and analyzes the possible reasons, providing time-saving information for guiding the researchers' studies.

In doped ZnO studies began to grow in number in 2010, though the first study was published on the Web of Science much earlier, in 1976. When the distribution of the number of articles using the keywords by year is shown, it can be seen that recent articles focus mainly on the synthesis of In doped ZnO film via different techniques such as sol-gel and the hydrothermal route, and on device applications such as solar cell and gas sensors. The distribution of the number of the articles by years is presented in Figure 4. The 2016 value in Figure 4 indicates the point of increase of the used-keyword number in articles up to the present. From the analysis of the publications on In doped ZnO, we found that IZO films are mostly used in diode applications, gas sensors and solar cells. Figure 4 shows the emergence of studies on nanoparticles, nanorods and gas sensors in the last few years, together with the advance of low-cost chemical routes such as hydrothermal, sol-gel and spray pyrolysis techniques. This allows us to predict the trends in future studies which will be focused on IZO films and nanostructures obtained by low-cost synthesis for gas sensor and optoelectronic applications.

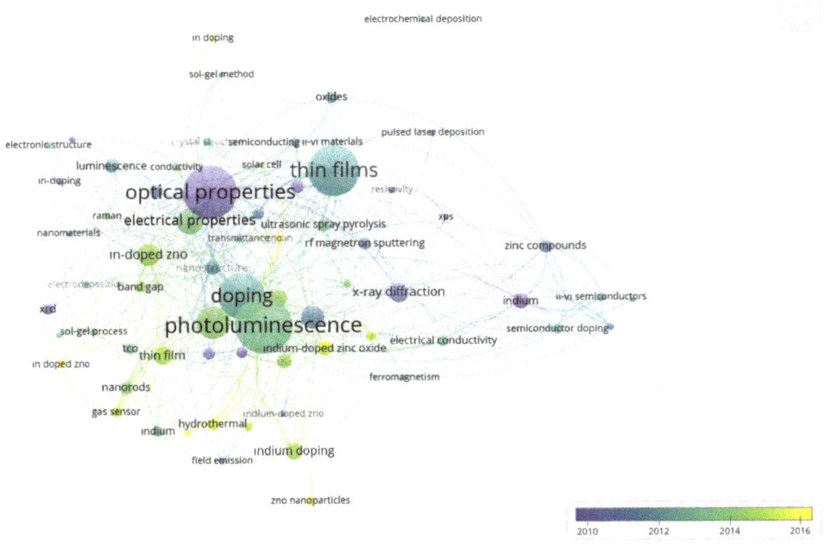


**Figure 4.** The distribution of the number of the articles by years.

In order to create a map based on text data for the most used words in the abstracts of articles, the Web of Science bibliographic database file was uploaded into the program. Subsequently, the abstract and binary counting method was selected as the field. The minimum number of occurrences of a term was set as 20 and the number of terms to be selected was automatically stated to be 98. The map created from this is reported in Figure 5.

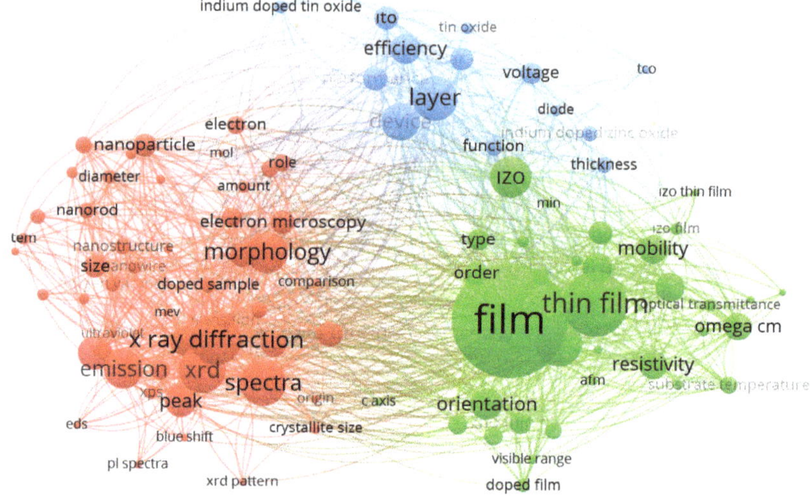


**Figure 5.** Most used words in abstract sections.

Figure 5 shows three clusters, and the words "film" (f = 327) and "thin film" (f = 185) are the most used words found in the abstracts. The most used words were also found to be: X-ray diffraction (f = 146), glass substrate (f = 135), X-ray Diffraction (XRD) (f = 135), spectra (f = 131) and layer (f = 128). From these results, it can be concluded that research on In doped ZnO generally focuses on its synthesis and characterization in its thin film form. In addition, the high number of words: glass, XRD, spectra and layer indicate that In-doped ZnO films are synthesized for use in different applications and XRD and spectral techniques are often used for their characterization. On the contrary, the frequency of the words "X-ray photoelectron spectroscopy (XPS), atomic force microscopy (AFM), Photoluminescence Spectroscopy (PL), energy dispersive x-ray spectroscopy (EDS)" was found to be less than the others. Therefore, it is recommended that researchers use these analyses in their studies. Considering that the size of the circles in Figure 5 depends on the frequency of the words in the abstract, it can be concluded that research on nano-shapes regarding In-doped ZnO is limited in this field. Additionally, the frequency of words such as "SEM"," electron microscopy", "nanostructure", "nanoparticle", "nanorod", "nanowire", "morphology", which indicate the analysis of different IZO morphology, was 71, 70, 46, 68, 47, 41 and 121, respectively. Nevertheless, the ZnO nanostructures have taken the lead in recent years thanks to advanced synthesis methods, and the number of publications about ZnO nanostructures is lower in comparison with those on thin films, and the relationship of the above mentioned words (SEM, nanoparticles, etc.) with the other keywords is weak, which explains their relevance observed in the charts. Therefore, it is important for researchers to study this subject to fill the gap in the field. Additionally, the words "device", "layer" and "thickness" indicated that In doped ZnO films have also been used in device applications as interlayer structures. Thus, it is recommended that the authors concentrate on studies investigating the performance of nano-shaped indium doped ZnO structures for device applications.

When the distribution of the most used words is shown on a year-by-year basis, it is evident that more recent articles tend to focus on the efficiency and performance of In doped ZnO films as well as TCO in device applications such as solar cells, diodes and photoluminescence applications. The distribution of the most used words in abstracts by year is presented in Figure 6. The 2014 value in Figure 6 indicates the point of increase of the used-words number in abstracts up to the present. Accordingly, the effect of nanoparticles, variation in crystallite size and morphology are the most examined features of In doped ZnO films in aforementioned device applications.

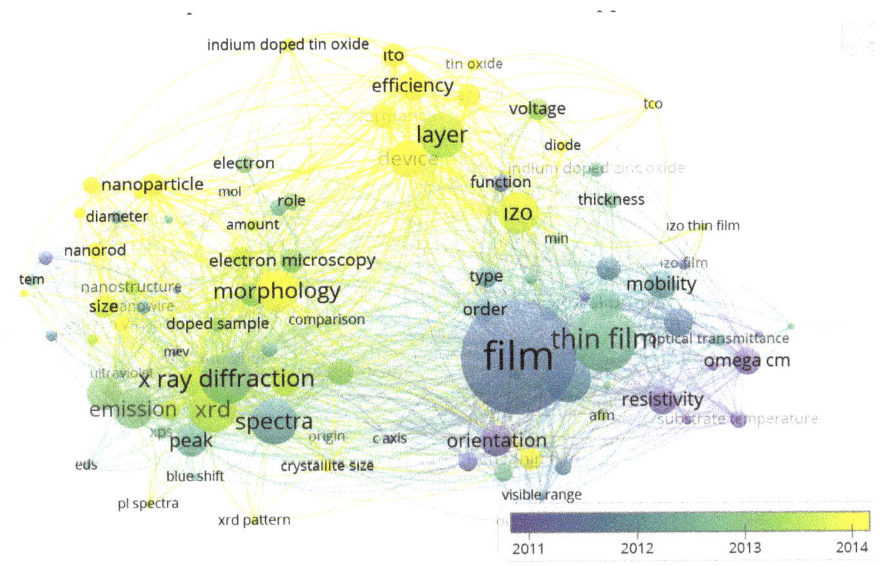

**Figure 6.** The distribution of the most used words in abstracts by year.

In order to create a map for most cited authors, citation analysis and authors were selected. The minimum number of documents by a particular author was set as five and the minimum number of citations of an author was determined as 10. The number of authors to be selected was automatically obtained as 25. The map created is shown in Figure 7. This indicates that Maldonado (384 citations) and Asomoza (286 citations) were the most cited authors in this field.

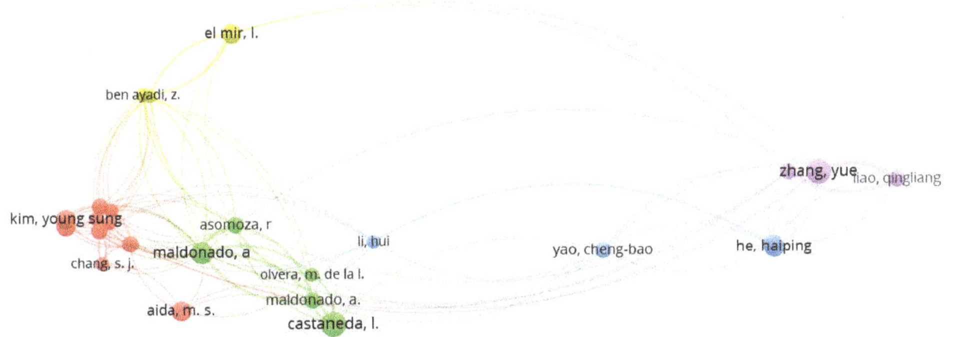

**Figure 7.** The most cited authors.

Examining the reference analysis of Maldonado's studies on Indium doped ZnO on Web of Science database, it is observed that articles titled "Indium-doped ZnO thin films deposited by the sol-gel technique [80]" and "Characterization of indium-doped zinc oxide films deposited by pyrolytic spray with different indium compounds as dopant [81]" have the most citations. Both studies were published in the "Thin Solid Film" journal. Researchers can use these studies as a source for their studies.

In addition, the co-citation analysis and cited authors were selected. The minimum number of citations of an author was set as 30 and the number of authors to be selected was automatically given as 44. The map created is shown in Figure 8. The co-citation analysis shows us the most cited authors in the reference list sections of the articles that were included in the analysis. It shows that Major (99 citations), Minami (87 citations) and Ozgur (85 citations) are most co-cited authors in this field. It can be said that these authors are leaders in the studies regarding In doped ZnO. Analyzing the studies of these authors will contribute to researchers' future studies on this subject.

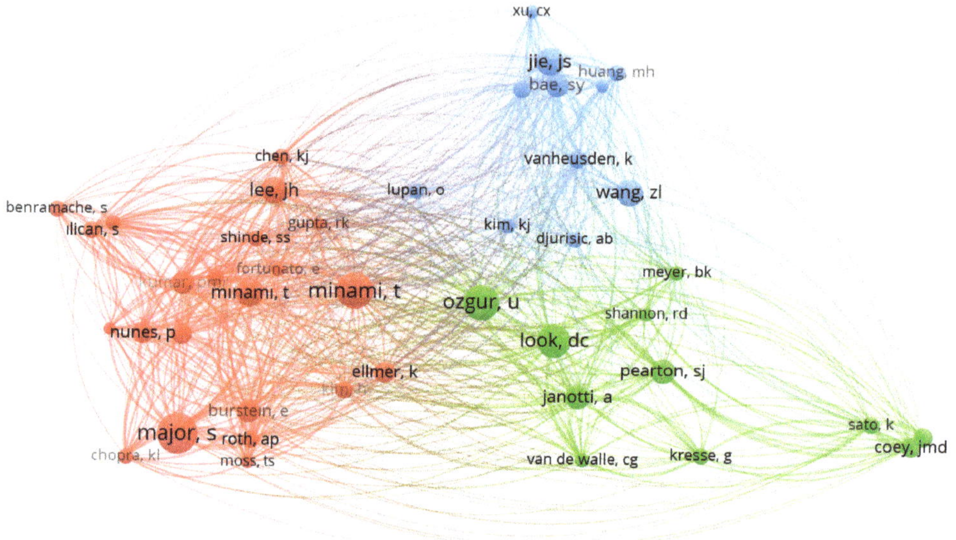


**Figure 8.** The most co-cited authors.

In order to create a map for most cited journals, the citation analysis and sources were selected. The minimum number of documents of a source was set as five and the minimum number of citations of a source was also set as five. The number of sources to be selected was automatically obtained as 43. The map created from this is presented in Figure 9. Figure 9 shows that the most cited journals are "Thin Solid Films (2434 citations, 49 documents), Sensors" and "Actuators B: Chemical (720 citations, 8 documents)" and "Applied Physics Letters (680 citations, 15 documents)".

Examining the reference analysis of "Thin Solid Films" on Indium doped ZnO on the Web of Science database, it is observed that articles titled "Zinc oxide thin films by the spray pyrolysis method [82]" and "Highly Transparent and Conducting Indium-Doped Zinc-Oxide Films by Spray Pyrolysis [83]" have the most citations. Both studies were published in the "Thin Solid Film" journal. Researchers can use these studies as a source in their future studies.

In addition to the citation analysis stated in the Figure 9, the number of articles in journals was also analyzed by years. Although the number of publications in "Thin Solid Film" is high, in recent years this journal has lost its popularity in regards to publications on In doped ZnO. It is seen that "Solar Energy", "Ceramics International", "Applied Physics-A", "Optik", "Material Research Express", "ACS Applied Materials and Interfaces", and "Optical Materials" published more studies since 2016 to the present. Therefore, it is suggested that these journals are preferred in future studies. The graph of all these results is presented in Figure 10.

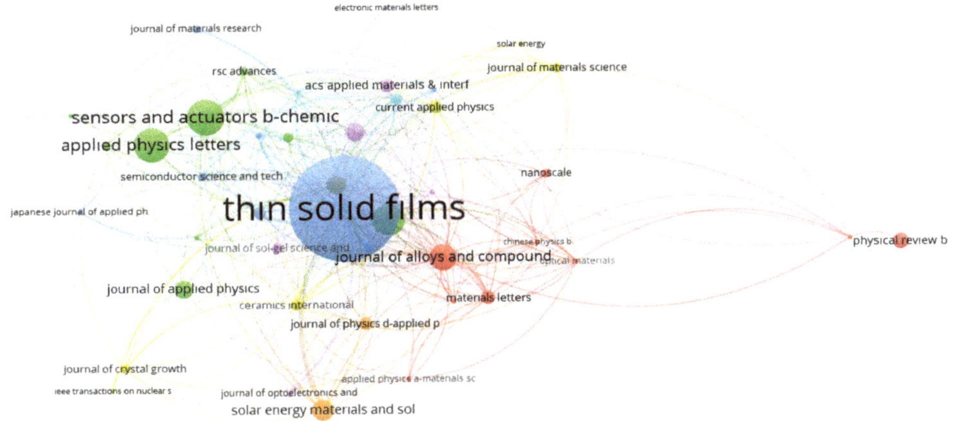

**Figure 9.** The most cited journals.

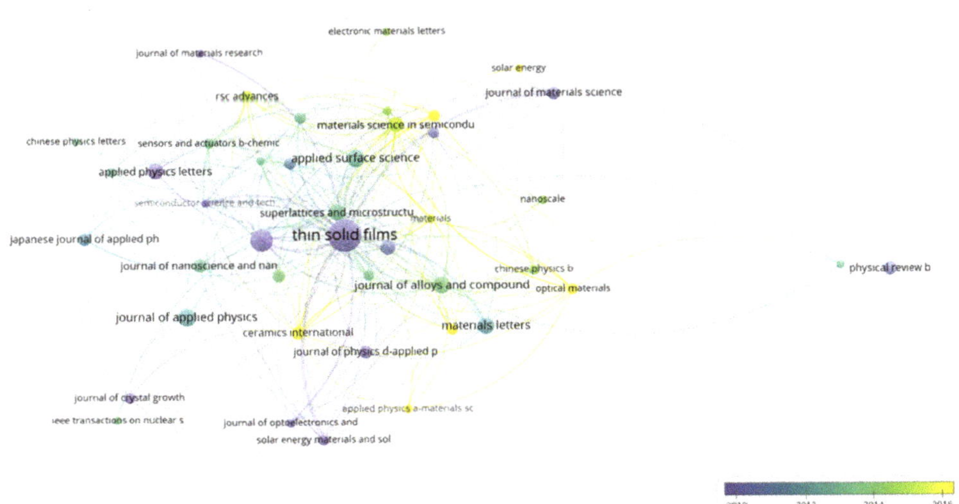

**Figure 10.** The journals with the most publications by years.

In addition, the co-citation analysis and cited sources were selected. The minimum number of citations of a source was set at 30 and the number of sources to be selected was automatically stated to 127. Figure 11 shows the resulting map. It shows that the most co-cited journals are "Applied Physics Letters" (2084 co-citations), "Thin Solid Films" (1624 co-citations), "Journal of Applied Physics" (1400 co-citations), "Physical Review B" (975 co-citations), and "Applied Surface Science" (626 co-citations). It is indicated that these journals are leaders in regards to their studies on In doped ZnO.

When the citation analysis of the studies on indium doped ZnOs on the Web of Science database in "Applied Physics Letters" is examined, it is seen that the studies named "Large and abrupt optical band gap variation in In-doped ZnO [84]" and "ZnO nanowires and nanobelts: Shape selection and thermodynamic modeling [85]" received more citations than the others. Authors may use these studies as important bibliographic sources.

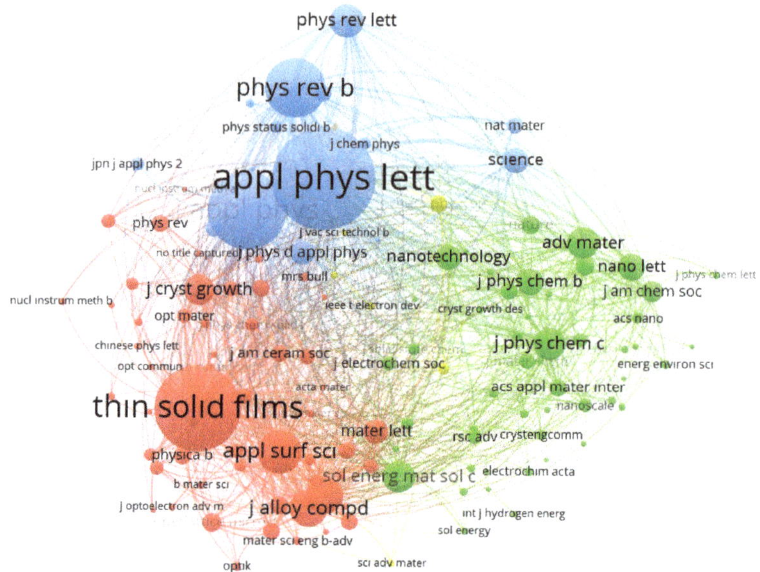


**Figure 11.** The most co-cited journals.

In order to create a map for co-authorship in countries, co-authorship and countries were selected. The minimum number of documents of a source of a country was adjusted as five and the minimum number of citations from a country was stated as five. The number of countries to be selected was automatically stated as 34. The created map is shown in Figure 12. Figure 12 shows that the countries with the higher number of documents are China (Citations = 4105, Documents = 177), India (Citations = 2104, Documents = 124), South Korea (Citations = 2268, Documents = 93), USA (Citations = 1573, Documents = 69), Taiwan (Citations = 681, Documents = 40) and Japan (Citations = 676, Documents = 39).

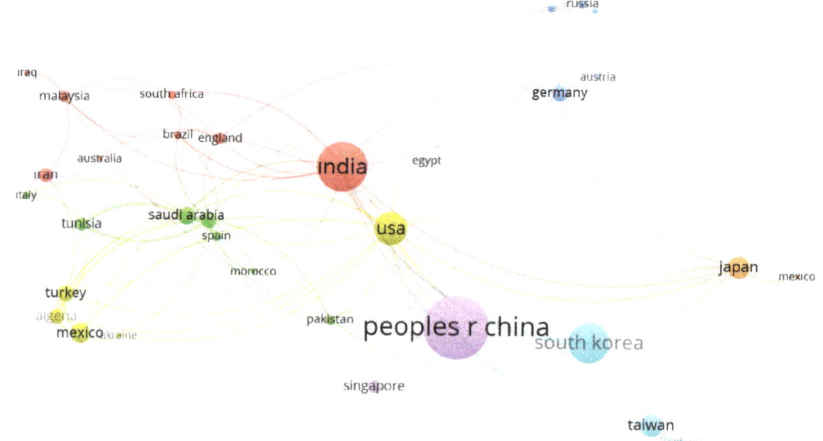


**Figure 12.** The co-authorship in countries.

According to the results, it is seen that China and India are the leading countries in this field, while in Mexico, Australia, Austria, Ukraine, Thailand, Egypt, Belgium, Netherlands, Iraq and Morocco, the studies on indium doped ZnO are very few. In this context, it is recommended that these countries collaborate with the leading countries in the field (China and India) to produce more publications.

## 4. Conclusions

In this study, a bibliometric analysis was conducted in order to reveal relationships between concepts in keywords and abstracts, the most cited/co-cited authors, the most cited/co-cited journals and the countries with the most academic studies in In doped ZnO studies. Results show that, apart from the obvious most used keywords such as ZnO and zinc oxide, photoluminescence, optical properties, thin films and doping are the other most used keywords. In addition, film, thin film, X-ray diffraction, glass substrate, XRD, and spectra are the most used words found in the abstracts. When the distribution of the words is searched on a year-by-year basis, it is evident that more recent articles tend to focus on efficiency and performance of In doped ZnO films as TCO in device applications such as solar cells, diodes, gas sensors and photoluminescence applications. The analysis of recent publications also shows an increasing interest towards IZO nanostructures (nanoparticles, nanorods, etc.) obtained by low cost chemical routes. Citation analysis reveals that Maldonado and Asomoza are the most cited authors, and Major, Minami and Ozgur are the most co-cited authors in this field. Thin Solid Films, Sensors and Actuators B: Chemical, Applied Physics Letters are the most cited journals and the most co-cited journals are Applied Physics Letters, Thin Solid Films, Journal of Applied Physics, Physical Review, Applied Surface Science. Finally, the countries with the highest number of documents are China, India, South Korea, USA and Japan, respectively. Based on all of our results and literature discussion, suggestions for authors are presented below:

- The studies mostly focus on some characteristics of In doped ZnO thin films such as structural, optical and electrical features. It is suggested that authors also pay attention to studies on the magnetic properties of ZnO. It is advised that the evaluation of ZnO's characteristics should also be made taking growth techniques into consideration.
- For researchers who want to examine the sensitivity of ZnO to ammonia, ethanol, acetone and water vapors, it may be suggested to work with (100) directed ZnO films due to the fact that films with (100) preferred orientation have much higher conductivity and higher sensing performance.
- Research on In doped ZnO generally focuses on its synthesis and characterization in its thin film form. However, studies on In-doped ZnO nanostructures are limited. Therefore, it is important for researchers to study this subject to fill the gap in the field.
- In doped ZnO films have been used in device applications as interlayer structures. Thus, it is recommended that the authors should concentrate on studies investigating the performance of non-shaped indium doped ZnO structures in device applications.
- Maldonado's studies on Indium doped ZnO entitled "Indium-doped ZnO thin films deposited by the sol-gel technique [63]" and "Characterization of indium-doped zinc oxide films deposited by pyrolytic spray with different indium compounds as dopant [64]" have the most citations. Researchers can use these studies and can also examine other works published in the same journal.
- The studies of Major, Minami and Ozgur could contribute to researchers' future studies on this subject. Researchers can use these studies as a relevant source in future studies.
- Articles entitled "Zinc oxide thin films by the spray pyrolysis method [65]" and "Highly Transparent and Conducting Indium-Doped Zinc-Oxide Films by Spray Pyrolysis [66]" have more citations than the others in Thin Solid Films. Researchers can use these studies as a source reference in future studies.
- Applied Physics Letters, Thin Solid Films, Journal of Applied Physics, Physical Review, Applied Surface Science are leader journals with their studies on IZO. However, Solar Energy, Ceramics International, Applied Physics-A, Optik, Material Research Express, ACS Applied Materials and Interfaces, Optical Materials published more studies from 2016 to the present. Therefore,

it is suggested that all these journals can contribute to the future study of the researchers on this subject.

- The studies entitled "Large and abrupt optical band gap variation in In-doped ZnO [67]" and "ZnO nanowires and nanobelts: Shape selection and thermodynamic modeling [68]" received more citations than others in the Applied Physics Letters. Authors could use these studies as a main source when writing a literature discussion in their future works.

- The countries with the highest number of documents on In doped ZnO are China, India, South Korea, USA and Japan. Researchers working in this field can undertake post-doctoral studies in the aforementioned countries. In addition, international projects can be produced by collaborating with these countries.

- In Mexico, Australia, Austria, Ukraine, Thailand, Egypt, Belgium, Netherlands, Iraq and Morocco, it is seen that the studies on indium doped ZnO are very few. It is recommended that these countries collaborate with the leading countries in the field (China and India) to produce more publications.

**Author Contributions:** Methodology, M.Y.; writing—original draft preparation, M.L.G.; writing—review and editing, G.T. All authors have read and agreed to the published version of the manuscript.

**Funding:** This research received no external funding.

**Acknowledgments:** The authors would like to thank Rabia Meryem Yilmaz from Ataturk University, Department of Computer Education and Instructional Technology for her valuable contributions to this study.

**Conflicts of Interest:** The authors declare no conflict of interest.

## References


1. Ranu, B.C.; Mandal, T. Indium(I) Iodide-Promoted Cleavage of Diaryl Diselenides and Disulfides and Subsequent Condensation with Alkyl or Acyl Halides. One-Pot Efficient Synthesis of Diorganyl Selenides, Sulfides, Selenoesters, and Thioesters. *J. Org. Chem.* **2004**, *69*, 5793–5795. [CrossRef] [PubMed]
2. Schneider, U.; Dao, H.T.; Kobayashi, S. Unusual Carbon–Carbon Bond Formations between Allylboronates and Acetals or Ketals Catalyzed by a Peculiar Indium(I) Lewis Acid. *Org. Lett.* **2010**, *12*, 2488–2491. [CrossRef] [PubMed]
3. Habashi, F. Indium, Physical and Chemical Properties. In *Encyclopedia of Metalloproteins*; Springer: New York, NY, USA, 2013; pp. 981–982.
4. Choubey, P.K.; Jha, M.K.; Gupta, D.; Jeong, J.; Lee, J.-C. Recovery of Rare Metal Indium (In) from Discarded LCD Monitors. In *Rare Metal Technology 2014*; John Wiley & Sons, Inc.: Hoboken, NJ, USA, 2014; pp. 37–42. ISBN 9781118888827.
5. Hofmann, M.; Hofmann, H.; Hagelüken, C.; Hool, A. Critical raw materials: A perspective from the materials science community. *Sustain. Mater. Technol.* **2018**, *17*, e00074. [CrossRef]
6. European Commission Committee and the Committee of the Regions on the 2017 list of Critical Raw Materials for the EU. *Commun. From Comm. Eur. Parliam. Counc. Eur. Econ. Soc. Comm. Comm. Reg.* **2017**, *XIV*, 293.
7. Yilmaz, M.; Grilli, M.L. The modification of the characteristics of nanocrystalline ZnO thin films by variation of Ta doping content. *Philos. Mag.* **2016**, *96*, 2125–2142. [CrossRef]
8. Yilmaz, M.; Tatar, D.; Sonmez, E.; Cirak, C.; Aydogan, S.; Gunturkun, R. Investigation of Structural, Morphological, Optical, and Electrical Properties of Al Doped ZnO Thin Films Via Spin Coating Technique. *Synth. React. Inorganic, Met. Nano-Metal Chem.* **2016**, *46*, 489–494. [CrossRef]
9. Yilmaz, M.; Turgut, G.; Aydin, S.; Ertugrul, M. Electrochemical deposition of ZnO thin films on to tin(IV) oxide:Fluorine. *Asian J. Chem.* **2012**, *24*, 3371–3374.
10. Sun, X.; Li, Q.; Jiang, J.; Mao, Y. Morphology-tunable synthesis of ZnO nanoforest and its photoelectrochemical performance. *Nanoscale* **2014**, *6*, 8769–8780. [CrossRef]
11. Kahraman, S.; Çakmak, H.M.; Çetinkaya, S.; Bayansal, F.; Çetinkara, H.A.; Güder, H.S. Characteristics of ZnO thin films doped by various elements. *J. Cryst. Growth* **2013**, *363*, 86–92. [CrossRef]
12. Nunes, P.; Braz Fernandes, F.M.; Silva, R.J.C.; Fortunato, E.; Martins, R. Structural Characterisation of Zinc Oxide Thin Films Produced by Spray Pyrolysis. *Key Eng. Mater.* **2002**, *230*, 599–602. [CrossRef]

13. Rashed, H.A.; Umran, N.M. The stability and electronic properties of Si-doped ZnO nanosheet: A DFT study. *Mater. Res. Express* **2019**, *6*, 045044. [CrossRef]

14. Habashyani, S.; Özmen, A.; Aydogan, S.; Yilmaz, M. An examination of correlation between characteristic and device performance of ZnO films as a function of La content. *Vacuum* **2018**, *157*, 497–507. [CrossRef]

15. Feng, T.-H.; Xia, X.-C. Characteristics of doping controllable ZnO films grown by photo-assisted metal organic chemical vapor deposition. *Opt. Mater. Express* **2017**, *7*, 1281. [CrossRef]

16. Sukee, A.; Kantarak, E.; Singjai, P. Preparation of Aluminum doped Zinc Oxide Thin Films on Glass Substrate by Sparking Process and Their Optical and Electrical Properties. *J. Phys. Conf. Ser.* **2017**, *901*, 012153. [CrossRef]

17. Arici, F.; Yildirim, P.; Caliklar, Ş.; Yilmaz, R.M. Research trends in the use of augmented reality in science education: Content and bibliometric mapping analysis. *Comput. Educ.* **2019**, *142*, 103647. [CrossRef]

18. Yilmaz, R.M.; Topu, F.B.; Takkaç Tulgar, A. An examination of the studies on foreign language teaching in pre-school education: A bibliometric mapping analysis. *Comput. Assist. Lang. Learn.* **2019**, 1–24. [CrossRef]

19. Liao, H.; Tang, M.; Luo, L.; Li, C.; Chiclana, F.; Zeng, X.-J. A Bibliometric Analysis and Visualization of Medical Big Data Research. *Sustainability* **2018**, *10*, 166. [CrossRef]

20. Melkers, J. Bibliometrics as a Tool for Analysis of R&D Impacts. In *Evaluating R&D Impacts: Methods and Practice*; Springer: Boston, MA, USA, 1993; pp. 43–61.

21. Silva-Lopez, H.E.; Marcelino, B.S.; Guillen-Cervantes, A.; Zelaya-Angel, O.; Ramirez-Bon, R. Physical Properties of Sputtered Indium-doped ZnO Films Deposited on Flexible Transparent Substrates. *Mater. Res.* **2018**, *21*, 1–6. [CrossRef]

22. Biswal, R.; Maldonado, A.; Vega-Pérez, J.; Acosta, D.R.; De La Luz Olvera, M. Indium Doped Zinc Oxide Thin Films Deposited by Ultrasonic Chemical Spray Technique, Starting from Zinc Acetylacetonate and Indium Chloride. *Materials* **2014**, *7*, 5038–5046. [CrossRef]

23. Du, X.; Liu, B.; Li, L.; Kong, X.; Zheng, C.; Lin, H.; Tong, Q.; Tao, S.; Zhang, X. Excimer emission induced intra-system self-absorption enhancement – a novel strategy to realize high efficiency and excellent stability ternary organic solar cells processed in green solvents. *J. Mater. Chem. A* **2018**, *6*, 23840–23855. [CrossRef]

24. Ting, C.-C.; Chang, S.-P.; Li, W.-Y.; Wang, C.-H. Enhanced performance of indium zinc oxide thin film transistor by yttrium doping. *Appl. Surf. Sci.* **2013**, *284*, 397–404. [CrossRef]

25. Moradi-Haji Jafan, M.; Zamani-Meymian, M.-R.; Rahimi, R.; Rabbani, M. The effect of solvents and the thickness on structural, optical and electrical properties of ITO thin films prepared by a sol–gel spin-coating process. *J. Nanostructure Chem.* **2014**, *4*, 89. [CrossRef]

26. Wade, K.; Banister, A. *The Chemistry of Aluminium, Gallium, Indium and Thallium: Comprehensive Inorganic Chemistry*; Pergamon Press: Oxford, UK, 2016.

27. Shaheera, M.; Girija, K.G.; Kaur, M.; Geetha, V.; Debnath, A.K.; Vatsa, R.K.; Muthe, K.P.; Gadkari, S.C. Characterization and device application of indium doped ZnO homojunction prepared by RF magnetron sputtering. *Opt. Mater.* **2020**, *101*, 109723. [CrossRef]

28. Yu, Y.; Yao, B.; He, Y.; Cao, B.; Ma, W.; Chang, L. Oxygen defect-rich In-doped ZnO nanostructure for enhanced visible light photocatalytic activity. *Mater. Chem. Phys.* **2020**, *244*, 122672. [CrossRef]

29. Bhatia, S.; Verma, N. Gas Sensing Performance of Dip-Coated Indium-Doped ZnO Films. *J. Electron. Mater.* **2018**, *47*, 6450–6457. [CrossRef]

30. Kyaw, A.K.K.; Wang, Y.; Zhao, D.W.; Huang, Z.H.; Zeng, X.T.; Sun, X.W. The properties of sol-gel processed indium-doped zinc oxide semiconductor film and its application in organic solar cells. *Phys. Status Solidi* **2011**, *208*, 2635–2642. [CrossRef]

31. El Filali, B.; Jaramillo Gomez, J.A.; Torchynska, T.V.; Casas Espinola, J.L.; Shcherbyna, L. Band-edge emission, defects, morphology and structure of in-doped ZnO nanocrystal films. *Opt. Mater.* **2019**, *89*, 322–328. [CrossRef]

32. Khalfallah, B.; Chaabouni, F.; Abaab, M. Some physical investigations on In-doped ZnO films prepared by RF magnetron sputtering using powder compacted target. *J. Mater. Sci. Mater. Electron.* **2015**, *26*, 5209–5216. [CrossRef]

33. Thambidurai, M.; Kim, J.Y.; Kang, C.M.; Muthukumarasamy, N.; Song, H.J.; Song, J.; Ko, Y.; Velauthapillai, D.; Lee, C. Enhanced photovoltaic performance of inverted organic solar cells with In-doped ZnO as an electron extraction layer. *Renew. Energy* **2014**, *66*, 433–442. [CrossRef]

34.	Benhaliliba, M.; Benouis, C.E.; Mouffak, Z.; Ocak, Y.S.; Tiburcio-Silver, A.; Aida, M.S.; Garcia, A.A.; Tavira, A.; Sanchez Juarez, A. Preparation and characterization of nanostructures of in-doped ZnO films deposited by chemically spray pyrolysis: Effect of substrate temperatures. *Superlattices Microstruct.* **2013**, *63*, 228–239. [CrossRef]

35.	Benouis, C.E.; Benhaliliba, M.; Sanchez Juarez, A.; Aida, M.S.; Chami, F.; Yakuphanoglu, F. The effect of indium doping on structural, electrical conductivity, photoconductivity and density of states properties of ZnO films. *J. Alloys Compd.* **2010**, *490*, 62–67. [CrossRef]

36.	Lee, C.; Lim, K.; Song, J. Highly textured ZnO thin films doped with indium prepared by the pyrosol method. *Sol. Energy Mater. Sol. Cells* **1996**, *43*, 37–45. [CrossRef]

37.	Rambu, A.P.; Sırbu, D.; Sandu, A.V.; Prodan, G.; Nıca, V. Influence of In doping on electro-optical properties of ZnO films. *Bull. Mater. Sci.* **2013**, *36*, 231–237. [CrossRef]

38.	Tang, K.; Gu, S.; Liu, J.; Ye, J.; Zhu, S.; Zheng, Y. Effects of indium doping on the crystallographic, morphological, electrical, and optical properties of highly crystalline ZnO films. *J. Alloys Compd.* **2015**, *653*, 643–648. [CrossRef]

39.	Jongthammanurak, S.; Cheawkul, T.; Witana, M. Morphological differences in transparent conductive indium-doped zinc oxide thin films deposited by ultrasonic spray pyrolysis. *Thin Solid Film.* **2014**, *571*, 114–120. [CrossRef]

40.	ILICAN, S.; İlgü Büyük, G. ZnO:Eu Filmlerinin Mikroyapısal ve Optik Özellikleri. *Karadeniz Fen Bilim. Derg.* **2018**, *8*, 141–153. [CrossRef]

41.	Marouf, S.; Beniaiche, A.; Guessas, H.; Azizi, A. Morphological, Structural and Optical Properties of ZnO Thin Films Deposited by Dip Coating Method. *Mater. Res.* **2016**, *20*, 88–95. [CrossRef]

42.	Karakaya, S. Annealing Effect on Structural and Optical Properties of ZnO Films Prepared by Ultrasonic Spray Pyrolysis. *ANADOLU Univ. J. Sci. Technol. A Appl. Sci. Eng.* **2016**, *17*, 670–676. [CrossRef]

43.	Purohit, A.; Chander, S.; Sharma, A.; Nehra, S.P.; Dhaka, M.S. Impact of low temperature annealing on structural, optical, electrical and morphological properties of ZnO thin films grown by RF sputtering for photovoltaic applications. *Opt. Mater.* **2015**, *49*, 51–58. [CrossRef]

44.	Ye, J.; Gu, S.; Zhu, S.; Chen, T.; Hu, L.; Qin, F.; Zhang, R.; Shi, Y.; Zheng, Y. The growth and annealing of single crystalline ZnO films by low-pressure MOCVD. *J. Cryst. Growth* **2002**, *243*, 151–156. [CrossRef]

45.	Prasada Rao, T.; Santhoshkumar, M.C. Highly oriented (100) ZnO thin films by spray pyrolysis. *Appl. Surf. Sci.* **2009**, *255*, 7212–7215. [CrossRef]

46.	Francombe, M.H.; Satō, H. *Single Crystal Films*; Pergamon Press: London, UK, 1964.

47.	Ilıcan, S.; Çağlar, Y.; Çağlar, M. X-ray Diffraction Studies of Undoped and in-Doped Cd0.22Zn0.78S Films Deposited by Spray Pyrolysis. *Cankaya Univ. J. Arts Sci.* **2005**, *1*, 85–94.

48.	Dimitrov, O.; Nesheva, D.; Blaskov, V.; Stambolova, I.; Vassilev, S.; Levi, Z.; Tonchev, V. Gas sensitive ZnO thin films with desired (002) or (100) orientation obtained by ultrasonic spray pyrolysis. *Mater. Chem. Phys.* **2014**, *148*, 712–719. [CrossRef]

49.	Kumar, V.; Singh, N.; Mehra, R.M.; Kapoor, A.; Purohit, L.P.; Swart, H.C. Role of film thickness on the properties of ZnO thin films grown by sol-gel method. *Thin Solid Films* **2013**, *539*, 161–165. [CrossRef]

50.	Liu, F.-C.; Li, J.-Y.; Chen, T.-H.; Chang, C.-H.; Lee, C.-T.; Hsiao, W.-H.; Liu, D.-S. Effect of Silver Dopants on the ZnO Thin Films Prepared by a Radio Frequency Magnetron Co-Sputtering System. *Materials* **2017**, *10*, 797. [CrossRef]

51.	Obeid, M.M.; Jappor, H.R.; Al-Marzoki, K.; Al-Hydary, I.A.; Edrees, S.J.; Shukur, M.M. Unraveling the effect of Gd doping on the structural, optical, and magnetic properties of ZnO based diluted magnetic semiconductor nanorods. *RSC Adv.* **2019**, *9*, 33207–33221. [CrossRef]

52.	Khashan, K.S.; Mahdi, M. Preparation of indium-doped zinc oxide nanoparticles by pulsed laser ablation in liquid technique and their characterization. *Appl. Nanosci.* **2017**, *7*, 589–596. [CrossRef]

53.	Hamberg, I.; Granqvist, C.G. Evaporated Sn-doped In$_2$O$_3$ films: Basic optical properties and applications to energy-efficient windows. *J. Appl. Phys.* **1986**, *60*, R123–R160. [CrossRef]

54.	Singh, A.; Chaudhary, S.; Pandya, D.K. High conductivity indium doped ZnO films by metal target reactive co-sputtering. *Acta Mater.* **2016**, *111*, 1–9. [CrossRef]

55.	Cao, Y.; Miao, L.; Tanemura, S.; Tanemura, M.; Kuno, Y.; Hayashi, Y.; Mori, Y. Optical Properties of Indium-Doped ZnO Films. *Jpn. J. Appl. Phys.* **2006**, *45*, 1623–1628. [CrossRef]

56. Aydoğan, Ş.; Grilli, M.L.; Yilmaz, M.; Çaldiran, Z.; Kaçuş, H. A facile growth of spray based ZnO films and device performance investigation for Schottky diodes: Determination of interface state density distribution. *J. Alloys Compd.* **2017**, *708*, 55–66. [CrossRef]

57. Lim, J.H.; Lee, S.M.; Kim, H.-S.; Kim, H.Y.; Park, J.; Jung, S.-B.; Park, G.C.; Kim, J.; Joo, J. Synergistic effect of Indium and Gallium co-doping on growth behavior and physical properties of hydrothermally grown ZnO nanorods. *Sci. Rep.* **2017**, *7*, 41992. [CrossRef] [PubMed]

58. Kim, Y.-S.; Tai, W.-P.; Shu, S.-J. Effect of preheating temperature on structural and optical properties of ZnO thin films by sol–gel process. *Thin Solid Films* **2005**, *491*, 153–160. [CrossRef]

59. Lim, S.Y.; Brahma, S.; Liu, C.-P.; Wang, R.-C.; Huang, J.-L. Effect of indium concentration on luminescence and electrical properties of indium doped ZnO nanowires. *Thin Solid Film.* **2013**, *549*, 165–171. [CrossRef]

60. Hori, Y.; Shiota, Y.; Ida, T.; Yoshizawa, K.; Mizuno, M. Local structures and electronic properties of In atoms in In-doped ZnO. *Thin Solid Film.* **2019**, *685*, 428–433. [CrossRef]

61. Chirakkara, S.; Nanda, K.K.; Krupanidhi, S.B. Pulsed laser deposited ZnO: In as transparent conducting oxide. *Thin Solid Films* **2011**, *519*, 3647–3652. [CrossRef]

62. Caglar, M.; Caglar, Y.; Ilican, S. Electrical and optical properties of undoped and In-doped ZnO thin films. *Phys. Status Solidi c* **2007**, *4*, 1337–1340. [CrossRef]

63. Kumar, P.M.R.; Kartha, C.S.; Vijayakumar, K.P.; Abe, T.; Kashiwaba, Y.; Singh, F.; Avasthi, D.K. On the properties of indium doped ZnO thin films. *Semicond. Sci. Technol.* **2004**, *20*, 120–126. [CrossRef]

64. Edinger, S.; Bansal, N.; Bauch, M.; Wibowo, R.A.; Újvári, G.; Hamid, R.; Trimmel, G.; Dimopoulos, T. Highly transparent and conductive indium-doped zinc oxide films deposited at low substrate temperature by spray pyrolysis from water-based solutions. *J. Mater. Sci.* **2017**, *52*, 8591–8602. [CrossRef]

65. Caglar, Y.; Zor, M.; Caglar, M.; Ilican, S. Influence of the indium incorporation on the structural and electrical properties of zinc oxide films. *J. Optoelectron. Adv. Mater.* **2006**, *8*, 1867–1873.

66. Nasir, M.F.; Hannas, M.; Mamat, M.H.; Rusop, M. Electrical Properties of Indium-Doped Zinc Oxide Nanostructures Doped at Different Dopant Concentrations. In *Proceedings of the Nanoscience, Nanotechnology and Nanoengineering: Fundamentals and Applications*; Trans Tech Publications Ltd.: Baech, Switzerland, 2015; Volume 1109, pp. 593–597.

67. Peng, L.; Fang, L.; Zhao, Y.; Wu, W.; Ruan, H.; Kong, C. Growth and characterization of indium doped zinc oxide films sputtered from powder targets. *J. Wuhan Univ. Technol. Sci. Ed.* **2017**, *32*, 866–870. [CrossRef]

68. Mahesh, D.; Kumar, M.C.S. Synergetic effects of aluminium and indium dopants in the physical properties of ZnO thin films via spray pyrolysis. *Superlattices Microstruct.* **2020**, *142*, 106511. [CrossRef]

69. Benzitouni, S.; Zaabat, M.; Mahdjoub, A.; Benaboud, A.; Boudine, B. High transparency and conductivity of heavily In-doped ZnO thin films deposited by dip-coating method. *Mater. Sci.* **2018**, *36*, 427–434. [CrossRef]

70. Ullah, M.; Chunlei, W.; Su, W.-B.; Manan, A.; Ahmad, A.S.; Rehman, A.U. Thermoelectric properties of indium-doped zinc oxide sintered in an argon atmosphere. *J. Mater. Sci. Mater. Electron.* **2019**, *30*, 4813–4818. [CrossRef]

71. Winkler, N.; Wibowo, A.; Kubicek, B.; Kautek, W.; Ligorio, G.; List-Kratochvil, E.; Dimopoulos, T. Rapid Processing of In-Doped ZnO by Spray Pyrolysis from Environment-Friendly Precursor Solutions. *Coatings* **2019**, *9*, 245. [CrossRef]

72. Mahesh, D.; Kumar, B.H.; Kumar, M.C.S. Enhanced luminescence property of 1 D nanorods realised by aqueous chemical growth on indium doped zinc oxide thin films. *Thin Solid Film.* **2019**, *686*, 137279. [CrossRef]

73. Kotlyarchuk, B.; Savchuk, V.; Oszwaldowski, M. Preparation of undoped and indium doped ZnO thin films by pulsed laser deposition method. *Cryst. Res. Technol.* **2005**, *40*, 1118–1123. [CrossRef]

74. Wang, B.; Callahan, M.J.; Xu, C.; Bouthillette, L.O.; Giles, N.C.; Bliss, D.F. Hydrothermal growth and characterization of indium-doped-conducting ZnO crystals. *J. Cryst. Growth* **2007**, *304*, 73–79. [CrossRef]

75. Ye, Z.-Z.; Tang, J.-F. Transparent conducting indium doped ZnO films by dc reactive S-gun magnetron sputtering. *Appl. Opt.* **1989**, *28*, 2817. [CrossRef]

76. Joseph, B.; Manoj, P.K.; Vaidyan, V.K. Studies on preparation and characterization of indium doped zinc oxide films by chemical spray deposition. *Bull. Mater. Sci.* **2005**, *28*, 487–493. [CrossRef]

77. Qin, G.-P.; Zhang, H.; Ruan, H.-B.; Wang, J.; Wang, D.; Kong, C.-Y. Effect of Post-Annealing on Structural and Electrical Properties of ZnO: In Films. *Chin. Phys. Lett.* **2019**, *36*, 047301. [CrossRef]

78. Barquinha, P.; Gonçalves, G.; Pereira, L.; Martins, R.; Fortunato, E. Effect of annealing temperature on the properties of IZO films and IZO based transparent TFTs. *Thin Solid Film.* **2007**, *515*, 8450–8454. [CrossRef]

79. Özmen, A.; Aydogan, S.; Yilmaz, M. Fabrication of spray derived nanostructured n-ZnO/p-Si heterojunction diode and investigation of its response to dark and light. *Ceram. Int.* **2019**, *45*, 14794–14805. [CrossRef]

80. Luna-Arredondo, E.J.; Maldonado, A.; Asomoza, R.; Acosta, D.R.; Meléndez-Lira, M.A.; Olvera, M. de la L. Indium-doped ZnO thin films deposited by the sol–gel technique. *Thin Solid Film.* **2005**, *490*, 132–136. [CrossRef]

81. Gómez, H.; Maldonado, A.; Asomoza, R.; Zironi, E.P.; Cañetas-Ortega, J.; Palacios-Gómez, J. Characterization of indium-doped zinc oxide films deposited by pyrolytic spray with different indium compounds as dopants. *Thin Solid Film.* **1997**, *293*, 117–123. [CrossRef]

82. Krunks, M.; Mellikov, E. Zinc oxide thin films by the spray pyrolysis method. *Thin Solid Films* **1995**, *270*, 33–36. [CrossRef]

83. Major, S.; Banerjee, A.; Chopra, K.L. Highly transparent and conducting indium-doped zinc oxide films by spray pyrolysis. *Thin Solid Film.* **1983**, *108*, 333–340. [CrossRef]

84. Kim, K.J.; Park, Y.R. Large and abrupt optical band gap variation in In-doped ZnO. *Appl. Phys. Lett.* **2001**, *78*, 475–477. [CrossRef]

85. Fan, H.J.; Barnard, A.S.; Zacharias, M. ZnO nanowires and nanobelts: Shape selection and thermodynamic modeling. *Appl. Phys. Lett.* **2007**, *90*, 143116. [CrossRef]

*Article*

# Study of Sputtered ITO Films on Flexible Invar Metal Foils for Curved Perovskite Solar Cells

**Hae-Jun Seok and Han-Ki Kim ***

School of Advanced Materials Science and Engineering, Sungkyunkwan University, Suwon, Gyeonggi-do 16419, Korea; hj.seok@skku.edu
* Correspondence: hankikim@skku.edu; Tel.: +82-31-290-7391

Received: 3 December 2018; Accepted: 22 January 2019; Published: 24 January 2019

**Abstract:** We have studied characteristics of tin-doped indium oxide (ITO) films sputtered on flexible invar metal foil covered with an insulating $SiO_2$ layer at room temperature to use as transparent electrodes coated substrates for curved perovskite solar cells. Sheet resistance, optical transmittance, surface morphology, and microstructure of the ITO films on $SiO_2$/invar substrate are investigated as a function of the thickness from 50 to 200 nm. The optimized ITO film exhibits a low sheet resistance of 50.21 Ohm/square and high optical transmittance of up to 94.31% even though it is prepared at room temperature. In particular, high reflectance of invar metal substrate could enhance the power conversion efficiency of curved perovskite solar cell fabricated on the $ITO/SiO_2$/invar substrate. In addition, critical bending radius of the 150 nm-thick ITO film is determined by lab-designed outer and inner bending tests to show feasibility as flexible electrode. Furthermore, dynamic fatigue test is carried out to show flexibility of the ITO film on invar metal substrate. This suggests that the $ITO/SiO_2$/invar substrate can be applied as flexible electrodes and substrates for curved perovskite solar cells.

**Keywords:** indium tin oxide (ITO); invar metal substrate; curved perovskite solar cells; flexibility; reflectance; electrode

---

## 1. Introduction

Methylammonium lead tri-iodide ($MAPbI_3$) perovskite solar cells (PSCs) have emerged as next generation photovoltaics following Si-based photovoltaics and organic photovoltaics due to their high power conversion efficiencies larger than 20% and solution-based simple fabrication process [1–5]. In particular, planar heterojunction type flexible perovskite solar cells have been extensively investigated because planar architecture is easily fabricated on flexible substrate at low temperature. However, most flexible PSCs fabricated on plastic substrate exhibited limit in reliability and stability due to high water vapor transmission rate (WVTR) values of conventional plastic substrates [6–11]. To solve the problem of typical plastic substrates, flexible metal foils with smooth surface have emerged as promising flexible substrate for flexible photovoltaics and flexible organic light emitting diodes (OLEDs) [12–16]. Especially, metal foil substrates have many advantages over polymer substrates, such as enhanced chemical stability and lower thermal expansion coefficients. Furthermore, a high temperature process is possible due to the higher thermal stability of metal foil compared to that of polymers, and a passivation layer is not required to prevent water vapor and oxygen migrating through the substrate [17–20]. On the other hand, impurities from the metal substrate and surface roughness are significant issues impeding their electronic applications. In particular, device performance may deteriorate due to out-diffusion of detrimental elements from the metal foil, such as Fe atoms, and electrical short circuits (i.e., direct contact of the anode and cathode electrode in solar cell devices) due to severe surface roughness [21–23]. Lee et al. suggested that invar metal foil is a promising flexible substrate material for flexible OLEDs [12]. Im et al. also reported that flexible PSCs fabricated on metal foil had a power conversion efficiency of 15% [24]. Typically, most of PSCs have been fabricated on physical vapor deposited Sn-doped $In_2O_3$ (ITO) electrode [25–27]. DC sputtered amorphous ITO films with fairly high sheet resistances (30–100 Ohm/square) and average optical transmittances of 80–85% have been widely employed as flexible electrodes for flexible optoelectronic devices. Although the potential of invar metal foil as flexible substrate for flexible optoelectronic devices has been well-known, detailed investigation of Sn-doped $In_2O_3$ (ITO) films on invar metal foil is still necessary. Recently, we reported high quality flexible ITO electrodes with a low sheet resistance of 15.75 Ohm/square, high optical transmittance of 85.88%, and outstanding flexibility grown by ion plating for flexible PSCs [28]. Because most research of PSCs has mainly been focused on active materials and interface buffer layers, investigation of flexible ITO films for curved or flexible PSCs is still lacking.

In this study, we investigated the electrical, optical, morphological and mechanical properties of DC sputtered ITO films on invar foil substrate as a function of ITO film thickness to show feasibility of invar metal substrate for curved perovskite solar cells. In particular, outer and inner bending tests were carried out to determine critical bending radius of ITO film on the invar metal substrate. In addition, resistance change of the ITO films during repeated bending cycles at a fixed bending radius of 15 mm was measured to show stable mechanical flexibility. The possible growth mechanism of ITO films on the $SiO_2$ coated invar metal substrate was suggested to correlate film thickness and characteristics of ITO films. In addition, we report a preliminary study of a perovskite solar cell fabricated on invar substrate.

## 2. Materials and Methods

The 120 μm thick invar metal foil (36% Ni–64% Fe) was employed as flexible substrate to deposit flexible ITO anode for curved perovskite solar cells. Before sputtering of ITO films, the 2 μm thick $SiO_2$ insulating layer was deposited on invar substrate using plasma enhanced chemical vapor deposition. Then, transparent ITO film with a thickness in the range 50–200 nm was sputtered on the $SiO_2$ coated invar substrate using a four-inch ITO target (Dasom RMS) at room temperature. During ITO sputtering process, DC power of 100 W, Ar flow rate of 20 sccm, and a working pressure of 3 mTorr was kept constant. Figure 1a shows the fabrication process of flexible ITO films on invar substrate. The ITO coated invar foil could be used as flexible substrate in curved perovskite solar cells as illustrated in Figure 1b. Due to opacity of invar substrate, the light was absorbed through the transparent top cathode unlike typical glass-based perovskite solar cells. The thickness of ITO films was measured using a profilometer (NanoMap LS, aep Technology, Newyork, NY, USA). Through Hall measurements (HMS-4000AM, Ecopia, Anyang-si, Korea) and a UV–visible spectrometer (UV 540, Unicam, Hachioji, Japan) analysis, the electrical and optical properties of the flexible ITO films were examined as a function of thickness. The morphology and microstructure of ITO film on $SiO_2$ coated invar substrate according to the increasing thickness was examined by field emission scanning electron microscope (FESEM: JSM-7600F, JEOL, Akishima, Japan) and X-ray diffraction (XRD, D8 ADVANCE, Bruker Corporation, Billerica, MA, USA). The mechanical flexibility of the ITO films on $SiO_2$ coated invar substrate was investigated by using a lab-designed outer/inner bending test machine (JUNIL TECH Co., DaeGu-Si, Korea). The resistance changes of the ITO films during outer and inner bending of substrate were measured with decreasing bending radius in order to determine critical bending radius. By pressing the clamped samples, we can control the bending radius. In outer and inner bending tests, both clamps moved simultaneously to specific position to make specific bending radius. Then, dynamic outer/inner fatigue tests were performed at a fixed bending radius of 15 mm for 10,000 cycles.

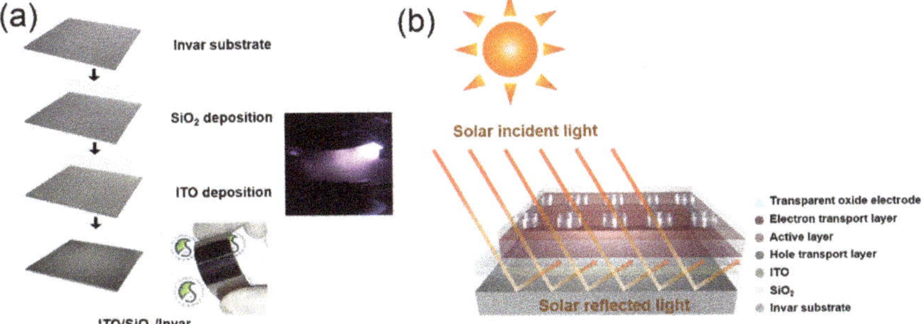

**Figure 1.** (**a**) Schematic fabrication process of flexible ITO films on SiO$_2$ coated invar metal substrate using typical DC magnetron sputtering at room temperature. The images show ITO sputtering process and curved invar substrate with ITO electrodes. (**b**) Device structure of curved perovskite solar cells with transparent InSnO anode and InZnSnO top cathode.

## 3. Results and Discussion

Figure 2a,b shows the Hall measurement results of ITO films sputtered on SiO$_2$ coated invar substrate and glass substrate in same chamber with increasing film thicness. With increasing ITO film thickness, the sheet resistance gradually decreased. Table 1 summarized sheet resistance of ITO films on the SiO$_2$ coated invar and glass substrate, respectively. Regardless of substrates, the DC sputtered ITO films showed a similar sheet resistance at a same film thickness. Because the insulating SiO$_2$ layer completely covered the invar substrate, the ITO film on SiO$_2$/invar substrate showed a similar sheet resistance to the ITO films on glass substrates. At a thickness of 200 nm, the ITO films on SiO$_2$/invar and glass substrate showed an identical sheet resistance of 44.57 and 44.52 Ohm/square and resistivity of $8.91 \times 10^{-4}$ and $8.90 \times 10^{-4}$ Ohm $\times$ cm, respectively.

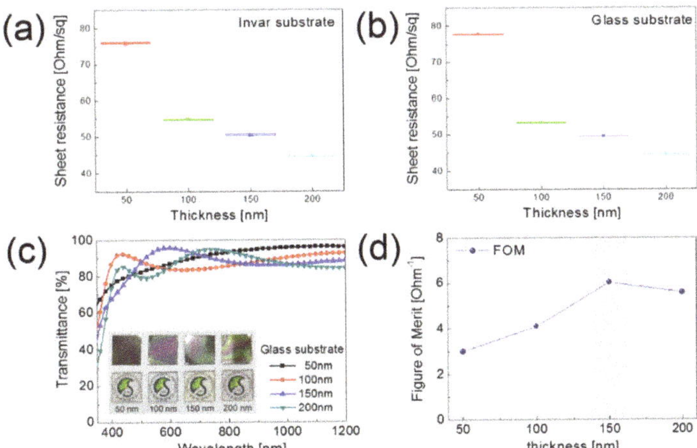

**Figure 2.** Hall measurement results of ITO films on (**a**) SiO$_2$ coated invar and (**b**) glass substrates prepared at room temperature with increasing thickness (50, 100, 150, 200 nm). (**c**) Optical transmittance of ITO films on glass substrates. Inset: pictures of ITO films deposited on invar substrates (upper figure) and on glass substrates (lower figure). (**d**) Figure of merit (FoM = $T^{10}/R_{sh}$) of ITO film on SiO$_2$/invar substrates as a function of thickness.

**Table 1.** Sheet resistance of ITO films on the $SiO_2$/invar and glass substrates with increasing ITO thickness.

| Substrate | Sheet Resistance (*Rs*) (Ohm/square) | | | |
|---|---|---|---|---|
| | 50 nm | 100 nm | 150 nm | 200 nm |
| Invar | 75.54 | 54.74 | 50.21 | 44.57 |
| Glass | 77.64 | 53.30 | 49.51 | 44.52 |

Figure 2c shows optical transmittance of ITO films on the glass substrate. Due to opacity of invar substrate, the optical transmittance of the ITO films was measured from the ITO film sputtered on glass substrate at same sputtering conditions with invar substrates. All ITO films show a high optical transmittance in the visible region regardless film thickness. With increasing ITO film thickness, oscillations in the optical transmittance was observed due to light interference phenomena [29,30]. Table 2 showed the average transmittance (400–800 nm) of ITO films on glass substrate as a function of thickness. At the 150nm-thick ITO film, the average transmittance had the highest value of 88.73%. To determine optimum ITO thickness, we compared figure of merit (FoM = $T^{10}/R_{sh}$) values, which were calculated from the average optical transmittance ($T$) and the sheet resistance ($R_{sh}$), as shown in Figure 2d [31]. Because the active layer of the PSCs absorbs visible light (400–800 nm), the FoM values are calculated using the average optical transmittance (400–800 nm). In addition, the voltage loss of the PSCs is influenced by the sheet resistance of the transparent electrode. Therefore, the exciton generation and voltage loss of the PSC are closely related to the FoM value, which is the quality of the transparent electrode. For those reasons, the optimum ITO thickness on the invar substrate was determined by the FoM value. Based on FoM calculation, we found that the 150 nm-thick ITO film had the highest FoM value of 6.02 Ohm$^{-1}$. In invar-based perovskite solar cells, optical reflectance on the surface of invar substrate is very important because photon penetrating perovskite active layer could be reflected from the surface of invar substrate as illustrated in Figure 1b.

**Table 2.** Average transmittance of ITO films on the glass substrate with increasing ITO thickness.

| Substrate | Average Transmittance (400–800 nm) (%) | | | |
|---|---|---|---|---|
| | 50 nm | 100 nm | 150 nm | 200 nm |
| Glass | 86.23 | 86.12 | 88.73 | 87.01 |

Figure 3a shows optical reflectance of ITO/$SiO_2$/invar samples with different ITO thickness. Due to high reflectance of invar substrate, all samples showed high reflectance regardless of ITO thickness. Optical reflectance of the ITO/$SiO_2$/invar samples are similar to $SiO_2$/invar substrate. Figure 3b exhibited XRD plots of ITO films on $SiO_2$/invar substrate with increasing thickness. The diffraction peaks appear at 2θ = 30.28°, 35.18°, 37.06°, 50.57°, and 60.15°, corresponding to the (222), (400), (411), (440), and (622) orientations. The several diffraction peaks demonstrate that the DC sputtered ITO films have a typical bixbyite structure even though they were prepared at room temperature. The intensity of XRD peaks increased due to well-developed polycrystalline structure and increase in sub-grain sizes. In addition, ITO film showed (222) preferred orientation with increasing film thickness. Kim et al. also reported that DC sputtered ITO films had a (222) preferred orientation with increasing film thickness [32]. Although we employed invar metal substrate, the ITO films ware directly sputtered on $SiO_2$ insulating layer, the microstructure of ITO film on $SiO_2$/invar substrate is similar to the ITO films on glass or fused silica ($SiO_2$) substrate.

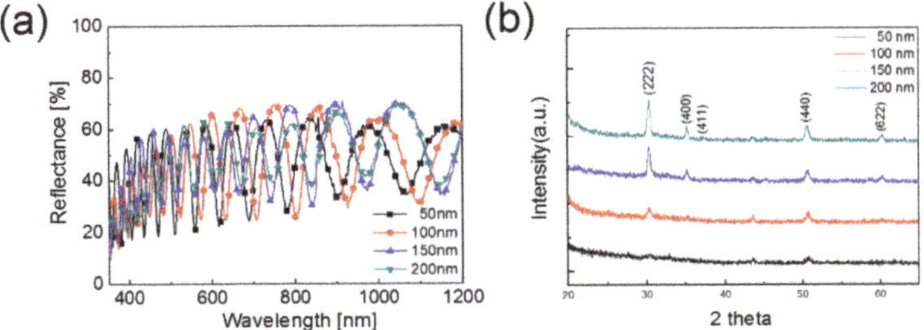

**Figure 3.** (**a**) Reflectance ITO films sputtered on SiO$_2$ coated invar substrate with increasing film thickness. (**b**) XRD plots of the ITO films sputtered on SiO$_2$ coated invar substrate at room temperature as a function of film thickness.

Figure 4 shows surface FESEM images of DC sputtered ITO films on SiO$_2$/invar substrate with increasing film thickness. At a thickness of 50 nm in Figure 4a, the ITO film showed garlic-flower like surface morphology due to amorphous sub-grain as expected from XRD plot. However, surface FESEM image of the ITO film with a thickness of 100 nm showed sub-grains consisting of micro-crystals as shown in Figure 4b. Further increase in thickness led to (222) preferred oriented surface morphology of ITO films as expected from XRD plots. As shown in Figure 4c, the 150-nm thick ITO films consisted of (222) oriented grains with a triangular-shape. The triangular-shaped ITO grains indicate that DC sputtered ITO films on SiO$_2$/invar substrate have a (222) preferred orientation. However, the 200-nm thick ITO films exhibit triangular-shaped (222), square-shaped (400), and rectangular-shaped (440) grains as shown in Figure 4d [27]. As expected from the XRD plots, these grain shapes indicate that the 200-nm thick ITO film consists of primarily (222), (400), and (440) grains. Due to the difference in sub-grain height, we can easily distinguish the sub-grain direction of DC-grown ITO films on SiO$_2$/invar substrate. As we reported in previous work [27], the surface of sputtered ITO films experienced severe ion bombardments from energetic particles in high density plasma. This bombardment resulted in re-sputtering of the specific plane of the ITO films. Due to different surface energy of special crystalline ITO planes, the re-sputtering phenomenon resulted in specific surface morphology of the ITO film depending on its preferred orientation.

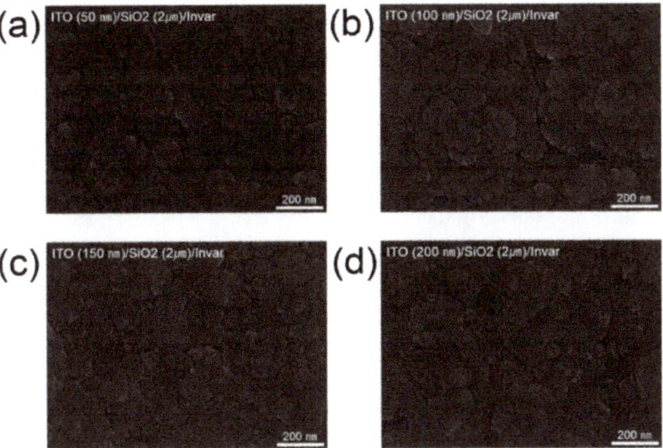


**Figure 4.** Surface FESEM images of DC sputtered ITO films on SiO$_2$/invar substrates; (**a**) 50 nm, (**b**) 100 nm, (**c**) 150 nm, and (**d**) 200 nm.

To determine the critical outer and inner bending radii of the ITO film on SiO$_2$/invar substrate, the resistance change ($\Delta R$) of the ITO film was measured in-situ by reducing the bending radius, as shown in Figure 5a. Here, the critical bending radius ($r_c$) is defined as the bending radius where the resistance abruptly increases due to the formation and propagation of cracks. The outer/inner bending test results showed that the 150 nm-thick ITO film on $SiO_2$/invar substrate had a constant resistance until the bending radii became 11 mm (outer bending) or 11 mm (inner bending), as shown in Figure 5a. Beyond this critical bending radius, the resistance of the film rapidly increased due to crack formation and physical separation of the ITO film. Due to thick 2 μm of $SiO_2$ on invar substrate, cracking in the ITO (150 nm)/$SiO_2$ multilayer led to increase in the resistance. Figure 5b shows the dynamic bending test results of the ITO (150 nm)/$SiO_2$/invar substrate with an increasing number of bending cycles at a fixed bending radius of 15 mm, which is larger than critical bending radius. The picture in upper panels in Figure 5b shows the dynamic outer and inner bending fatigue test steps with increasing bending cycles, up to 10,000 cycles.

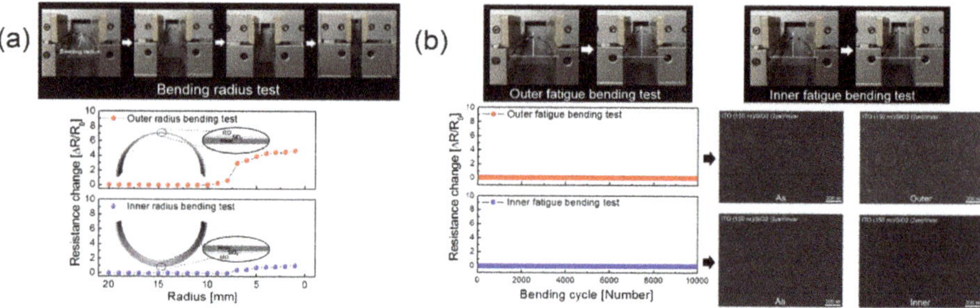

**Figure 5.** (**a**) Upper panels show pictures of the outer bending test steps with decreasing bending radius. Results of the outer and inner bending test with decreasing bending radius of 150 nm thick ITO films on $SiO_2$/invar substrate. Insets show the schematics of curved sample experiencing tensile and compressive stress. (**b**) Results of the outer and inner dynamic bending fatigue tests at a fixed bending radius of 15 mm for the 150 nm thick ITO films on $SiO_2$/invar substrate. The right side shows FESEM images of the ITO film before and after dynamic fatigue test after 10,000 cycles.

Figure 5b shows both dynamic outer and inner bending fatigue tests of the ITO (150 nm)/$SiO_2$/invar substrate; no change in the resistance ($\Delta R$) is observed, even after 10,000 bending cycles, demonstrating the outstanding flexibility and mechanical reliability of the ITO (150 nm)/$SiO_2$/invar substrate. Even after 10,000 outer and inner bending cycles, the surface of the ITO film showed an identical morphology to the as-deposited ITO film. The crystalline and featureless surface morphology of the sample is typical of a crystalline ITO film prepared at room temperature. Considering the application of curvature of perovskite solar cells, the outstanding flexibility and small critical bending radius of ITO film on $SiO_2$/invar substrate is acceptable in fabrication of curved perovskite solar cells.

Based on FoM value and bending test results, curved PSC was fabricated on a 150 nm-thick ITO film. Figure 6a depicts current density-voltage ($J$-$V$) curve of the PSC on the ITO (150 nm)/$SiO_2$/invar substrate. Figure 6b shows schematic representation of the PSC fabrication processes on the ITO (150 nm)/$SiO_2$/invar substrate. The ITO (150 nm)/$SiO_2$/invar substrate underwent UV/ozone treatment for 10 min before the deposition of a NiO layer. The desired amount of ethanolic precursor solution containing nickel diacetate tetra-hydrate ($Ni(CH_3COO)_2 \cdot 4H_2O$) and HCl was refluxed at 60 °C for 2 h. Then, NiO layer with an average thickness of 12 nm was fabricated by spin-coating at 3000 rpm for 40 s in air and the film was annealed at 100 °C for 30 min. Subsequently, the $MAPbI_3$ was coated onto the NiO layer by spin-coating at 4000 rpm for 45 s in air. The electron transport layer PCBM was grown on the $MAPbI_3$ layer by spin coating at 1000 rpm for 60 s and the film 100 °C for 10 min. Finally, 200 nm IZTO cathodes were formed successively, with the damage-free linear facing target sputtering by using a dumbbell-shaped shadow mask [33]. Table 3 summarizes the key $J$-$V$ parameters of the curved PSC. The $J$-$V$ curve of the performing device yields a $V_{OC}$ of 1.11 V, a $J_{SC}$ of

18.12 mA/cm$^2$, a fill factor of 47%, and a PCE of 9.70%. Although the PCE of the curved PSC fabricated on ITO/SiO$_2$/invar substrate is lower than glass based PSC at this moment, we strongly believe that further optimization of fabrication process could increase PCE of the curved PSC on invar substrate. However, successful operation of curved PSC on invar substrate demonstrates the feasibility of flexible invar substrate for next-generation curved PSCs.

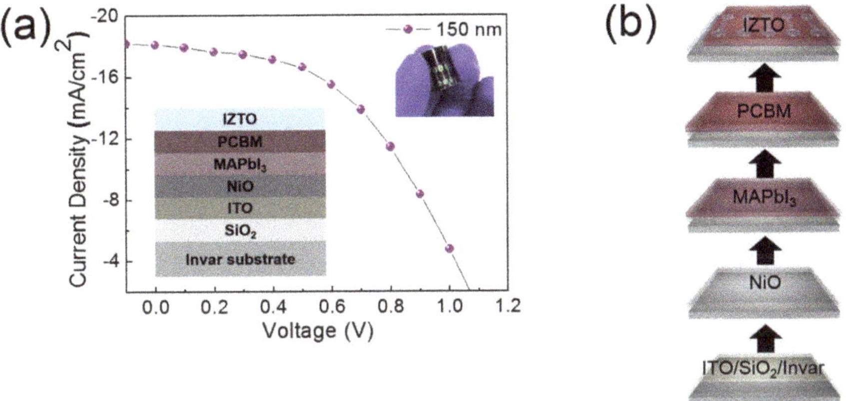


**Figure 6.** (a) Current density-voltage (*J*-*V*) curve of PSC fabrication on the ITO (150 nm)/SiO$_2$/invar substrate (under AM 1.5 solar light). (b) Schematic of a PSC fabrication procedure on the ITO (150 nm)/SiO$_2$/invar substrate.

**Table 3.** Key parameters of PSC based on ITO (150 nm)/SiO$_2$/invar substrate.

| ITO Thickness (nm) | $V_{OC}$ (V) | $J_{SC}$ (mA/cm$^2$) | Fill Factor (FF, %) | Power Conversion Efficiency (PCE, %) |
|---|---|---|---|---|
| 150 | 1.11 | 18.12 | 47.0 | 9.70 |

## 4. Conclusions

The electrical, optical, morphological, and structural properties of DC sputtered ITO films on SiO$_2$/invar substrates were investigated as a function of ITO thickness. We found that the resistivity and optical transmittance of DC sputtered ITO film on SiO$_2$/invar is dependent on film thickness. With increasing thickness, the ITO film showed a well-developed crystalline structure with (222) preferred orientation. Based on outer/inner bending test and dynamic fatigue tests, we confirmed the outstanding mechanical flexibility of the ITO/SiO$_2$/invar substrate. Based on electrical, optical, and mechanical properties of ITO film on SiO$_2$/invar substrate, we suggested that the DC sputtered ITO film on SiO$_2$/invar substrate could substitute ITO film on plastic substrate for curved perovskite solar cells. In addition, a preliminary study of the ITO coated-invar metal foils as a flexible substrate for curved PSCs indicates that invar metal foils are promising flexible substrates to substitute typical flexible polymer substrates for high performance curved PSCs.

**Author Contributions:** Design of experiment, H.-K.K.; Sputtering and Analysis, H.-J.S.

**Funding:** This work was supported by the Korea Institute of Energy Technology Evaluation and Planning (KETEP) and the Ministry of Trade, Industry & Energy (MOTIE) of the Republic of Korea (No. 20163010012200). This study also received partial support from Korea Electric Power Corporation (KEPCO, CX72170049).

**Conflicts of Interest:** The authors declare no conflict of interest.

## References


1. Jamal, M.S.; Bashar, M.S.; Mahmud Hasan, A.K.; Almutairi, Z.A.; Alharbi, H.F.; Alharthi, N.H.; Karim, M.R.; Misran, H.; Amin, N.; Sopian, K.B.; et al. Fabrication techniques and morphological analysis of perovskite absorber layer for high-efficiency perovskite solar cell: A review. *Renew. Sustain. Energy Rev.* **2018**, *98*, 469–488. [CrossRef]

2. Stranks, S.D.; Snaith, H.J. Metal-halide perovskites for photovoltaic and light-emitting devices. *Nat. Nanotechnol.* **2015**, *10*, 391–402. [CrossRef]

3. Li, J.; Jiu, T.; Duan, C.; Wang, Y.; Zhang, H.; Jian, H.; Zhao, Y.; Wang, N.; Huang, C.; Li, Y. Improved electron transport in MAPbI$_3$ perovskite solar cells based on dual doping graphdiyne. *Nano Energy* **2018**, *46*, 331–337. [CrossRef]

4. Saliba, M.; Matsui, T.; Seo, J.Y.; Domanski, K.; Correa-Baena, J.P.; Nazeeruddin, M.K.; Zakeeruddin, S.M.; Tress, W.; Abate, A.; Hagfeldt, A.; Grätzel, M. Cesium-containing triple cation perovskite solar cells: Improved stability, reproducibility and high efficiency. *Energy Environ. Sci.* **2016**, *9*, 1989–1997. [CrossRef] [PubMed]

5. You, S.; Wang, H.; Bi, S.; Zhou, J.; Qin, L.; Qiu, X.; Zhao, Z.; Xu, Y.; Zhang, Y.; Shi, X.; Zhou, H.; Zhiyong, T. A biopolymer heparin sodium interlayer anchoring TiO$_2$ and MAPbI$_3$ enhances trap passivation and device stability in perovskite solar cells. *Adv. Mater.* **2018**, *30*, 1706924. [CrossRef]

6. Bu, T.; Li, J.; Zheng, F.; Chen, W.; Wen, X.; Ku, Z.; Peng, Y.; Zhong, J.; Cheng, Y.B.; Huang, F. Universal passivation strategy to slot-die printed SnO$_2$ for hysteresis-free efficient flexible perovskite solar module. *Nat. Commun.* **2018**, *9*, 4609. [CrossRef] [PubMed]

7. Asghar, M.I.; Zhang, J.; Wang, H.; Lund, P.D. Device stability of perovskite solar cells—A review. *Renew. Sustain. Energy Rev.* **2017**, *77*, 131–146. [CrossRef]

8. Giacomo, F.D.; Fakharuddin, A.; Jose, R.; Brown, T.M. Progress, challenges and perspectives in flexible perovskite solar cells. *Energy Environ. Sci.* **2016**, *9*, 3007–3035. [CrossRef]

9. Shin, S.S.; Yang, W.S.; Noh, J.H.; Suk, J.H.; Jeon, N.J.; Park, J.H.; Kim, J.S.; Seong, W.M.; Seok, S.I. High-performance flexible perovskite solar cells exploiting Zn$_2$SnO$_4$ prepared in solution below 100 °C. *Nat. Commun.* **2015**, *6*, 7410. [CrossRef] [PubMed]

10. Zardetto, V.; Brown, T.M.; Reale, A.; Carlo, A.D. Substrates for flexible electronics: A practical investigation on the electrical, film flexibility, optical temperature, and solvent resistance properties. *J. Polym. Sci. Part B Polym. Phys.* **2011**, *49*, 638–648. [CrossRef]

11. Burrows, P.E.; Graff, G.L.; Gross, M.E.; Martin, P.M.; Shi, M.K.; Hall, M.; Mast, E.; Bonham, C.; Bennett, W.; Sullivan, M.B. Ultra barrier flexible substrates for flat panel displays. *Displays* **2001**, *22*, 65–69. [CrossRef]

12. Kim, K.; Kim, S.; Jung, G.H.; Lee, I.; Kim, S.; Ham, J.; Dong, W.J.; Hong, K.; Lee, J.L. Extremely flat metal films implemented by surface roughness transfer for flexible electronics. *RSC Adv.* **2018**, *8*, 10883–10888. [CrossRef]

13. Luo, Z.; Lu, Y.; Singer, D.W.; Berck, M.E.; Somers, L.A.; Goldsmith, B.R.; Charlie Johnson, A.T. Effect of substrate roughness and feedstock concentration on growth of wafer-scale graphene at atmospheric pressure. *Chem. Mater.* **2011**, *23*, 1441–1447. [CrossRef]

14. Calabrese, G.; Pettersen, S.V.; Pfüller, C.; Ramsteiner, M.; Grepstad, J.K.; Brandt, O.; Geelhaar, L.; Fernandez-Garrido, S. Effect of surface roughness, chemical composition, and native oxide crystallinity of the orientation of self-assembled GaN nanowires on Ti foils. *Nanotechnology* **2017**, *28*, 425602. [CrossRef]

15. Hong, K.; Yu, H.K.; Lee, I.; Kim, S.; Kim, Y.; Kim, K.; Lee, J.L. Flexible top-emitting organic light emitting diodes with a functional dielectric reflector on a metal foil substrate. *RSC Adv.* **2018**, *8*, 26156–26160. [CrossRef]

16. Cheon, J.H.; Choi, J.H.; Hur, J.H.; Jang, J.; Shin, H.S.; Jeong, J.K.; Mo, Y.G.; Chung, H.K. Active-matrix OLED on bendable metal foil. *IEEE Trans. Electron Devices* **2006**, *53*, 1273–1276. [CrossRef]

17. Kim, M.; Kim, K.B.; Jeon, C.W.; Lee, D.; Lee, S.N.; Lee, J.M.; Lee, H.C. CIGS solar cell devices on steel substrates coated with Na containing AIPO$_4$. *J. Phys. Chem. Solids* **2015**, *86*, 223–228. [CrossRef]

18. Kim, K.B.; Kim, M.; Lee, H.C.; Park, S.W.; Jeon, C.W. Copper indium gallium selenide (CIGS) solar cell devices on steel substrates coated with thick SiO$_2$-based insulating material. *Mater. Res. Bull.* **2017**, *85*, 168–175. [CrossRef]

19. Kim, K.B.; Kim, M.; Baek, J.; Park, Y.J.; Lee, J.R.; Kim, J.S.; Jeon, C. Influence of Cr thin films on the properties of flexible CIGS solar cells on steel substrates. *Electron. Mater. Lett.* **2014**, *10*, 247–251. [CrossRef]

20. Si, P.Z.; Choi, C.J. High hardness nanocrystalline invar alloys prepared from Fe-Ni nanoparticles. *Metals* **2018**, *8*, 28. [CrossRef]

21. Herz, K.; Eicke, A.; Kessler, F.; Wächter, R.; Powalla, M. Diffusion barrier for CIGS solar cells on metallic substrates. *Thin Solid Films* **2003**, *431–432*, 392–397. [CrossRef]

22. Batchelor, W.K.; Repins, I.L.; Schaefer, J.; Beck, M.E. Impact of substrate roughness on $CuIn_xGa_{1-x}Se_2$ device properties. *Sol. Energy Mater. Sol. Cells* **2004**, *83*, 67–80. [CrossRef]

23. Platzer-Björkman, C.; Jani, S.; Westlinder, J.; Linnarsson, M.K.; Scragg, J.; Edoff, M. Diffusion of Fe and Na in co-evaporated $Cu(In,Ga)Se_2$ devices on steel substrates. *Thin Solid Films* **2013**, *535*, 188–192. [CrossRef]

24. Heo, J.H.; Shin, D.H.; Lee, M.L.; Kang, M.G.; Im, S.H. Efficient organic-inorganic hybrid flexible perovskite solar cells prepared by lamination of polytriarylamine/$CH_3NH_3PbI_3$/anodized Ti metal substrate and graphene/PDMS transparent electrode substrate. *ACS Appl. Mater. Interfaces* **2018**, *10*, 31413–31421. [CrossRef] [PubMed]

25. Suyama, T.; Bae, H.; Setaka, K.; Ogawa, H.; Fukuoka, Y.; Suzuki, H.; Toyoda, H. Quantitative evaluation of high-energy O-ion particle flux in a DC magnetron sputter plasma with an indium-tin-oxide target. *J. Phys. D Appl. Phys.* **2017**, *50*, 445201. [CrossRef]

26. Lee, K.Y.; Chen, L.C.; Wu, Y.J. Effect of oblique-angle sputtered ITO electrode in $MAPbI_3$ perovskite solar cell structures. *Nanoscale Res. Lett.* **2017**, *12*, 556. [CrossRef]

27. Kim, J.H.; Seong, T.Y.; Ahn, K.J.; Chung, K.B.; Seok, H.J.; Seo, H.J.; Kim, H.K. The effects of film thickness on the electrical, optical, and structural properties of cylindrical, rotating, magnetron-sputtering ITO films. *Appl. Surf. Sci.* **2018**, *440*, 1211–1218. [CrossRef]

28. Kim, J.H.; Seok, H.J.; Seo, H.J.; Seong, T.Y.; Heo, J.H.; Lim, S.H.; Ahn, K.J.; Kim, H.K. Flexible ITO films with atomically flat surfaces for high performance flexible perovskite solar cells. *Nanoscale* **2018**, *10*, 20587–20598. [CrossRef]

29. Akman, E.; Cerkezoglu, E. Compositional and micro-scratch analyses of laser induced colored surface of titanium. *Opt. Lasers Eng.* **2016**, *84*, 37–43. [CrossRef]

30. Pérez del Pino, A.; Fernández-Pradas, J.M.; Serra, P.; Morenza, J.L. Coloring of titanium through laser oxidation: Comparative study with anodizing. *Surf. Coat. Technol.* **2004**, *187*, 106–112. [CrossRef]

31. Haacke, G. New figure of merit for transparent conductors. *J. Appl. Phys.* **1976**, *47*, 4086–4089. [CrossRef]

32. Kim, H.; Horwitz, J.S.; Kushto, G.; Piqué, A.; Kafafi, Z.H.; Gilmore, C.M.; Chrisey, D.B. Effect of film thickness on the properties of indium tin oxide thin films. *J. Appl. Phys.* **2000**, *88*, 6021–6025. [CrossRef]

33. Lee, J.H.; Shin, H.S.; Na, S.I.; Kim, H.K. Transparent and flexible PEDOT:PSS electrodes passivated by thin IZTO film using plasma-damage free linear facing target sputtering for flexible organic solar cells. *Sol. Energy Mater. Sol. Cells* **2013**, *109*, 192–198. [CrossRef]

*metals*

*Article*

# Suppressing the Use of Critical Raw Materials in Joining of AISI 304 Stainless Steel Using Activated Tungsten Inert Gas Welding

Sebastian Balos [1,*], Miroslav Dramicanin [1], Petar Janjatovic [1], Ivan Zabunov [2], Branka Pilic [3], Saurav Goel [4] and Magdalena Szutkowska [5]

1   Faculty of Technical Sciences, University of Novi Sad, Trg Dositeja Obradovica 6, 21000 Novi Sad, Serbia; dramicanin@uns.ac.rs (M.D.); janjatovic@uns.ac.rs (P.J.)
2   Faculty of Special Technology, Alexander Dubček University of Trenčín, Študentská 2, 911 50 Trenčín, Slovakia; ivan.zabunov@tnuni.sk
3   Faculty of Technology, University of Novi Sad, Bulevar Cara Lazara 1, 21000 Novi Sad, Serbia; brapi@uns.ac.rs
4   School of Aerospace, Transport and Manufacturing, Cranfield University, Bedfordshire MK43 0AL, UK; saurav.goel@cranfield.ac.uk
5   Łukasiewicz Research Network-Institute of Advanced Manufacturing Technology, Wroclawska 37a, 30-011 Krakow, Poland; szutkows@ios.krakow.pl
*   Correspondence: sebab@uns.ac.rs; Tel.: +381-21-485-2339

Received: 23 September 2019; Accepted: 21 October 2019; Published: 4 November 2019

**Abstract:** The aim of this study was to study the influence of $TiO_2$ coating for its efficacy during the activated-tungsten inert gas (TIG) welding and to suppress the use of consumables that are rich in critical raw materials. Post-welding penetration depth, particle size distribution, microstructure, and microhardness of welded samples were assessed. Based on these results, it was found that there is no direct correlation between the weld metal surface area and the coating. The particle size in the coating, although, seemed to have played an important role, e.g., nanoparticles resulted in an increased penetration depth and depth/width (D/W) ratio as opposed to the submicron-sized particles. The most optimal welding condition resulted when a mixture of submicron-sized and nanometric-sized particles were used. It was demonstrated by the Zeta analyser results that the micron particles rub the nanoparticles due to mechanical friction resulting in smaller oxide particle formation in the coating. Finally, the presence of Marangoni convection in TIG and reversed Marangoni convection in the activated TIG (A-TIG) process were proven by means of the microstructure analysis and measurement, which were found to be positively correlated.

**Keywords:** A-TIG welding; particle size; metal flow; penetration depth

## 1. Introduction

Tungsten inert gas (TIG), alternatively called gas tungsten arc welding (GTAW), is a well-established welding process that can produce high-quality welds on different materials, including stainless steels and a wide variety of non-ferrous alloys. However, as opposed to gas metal arc welding (GMAW), the process suffers from a relatively low yield so the application of TIG is traditionally limited to relatively thin sections in different welding positions [1,2].

To address the problem of yield, activated TIG (A-TIG), which uses a coating or a flux to act as a catalyser during the welding process (catalyzed TIG welding), was developed. The application of coating before TIG welding was proposed for the first time by the Paton Welding Institute of the National Academy of Sciences, Ukraine, back in the 1960s [3]. During A-TIG, a coating is sprayed or applied by a brush over the previously cleaned and prepared surface to be welded. Coatings are usually fabricated by mixing metallic oxide powders with solvents, most frequently acetone and ethanol [4–8].

The A-TIG process offers increased penetration depth, offering the possibility to weld significantly higher thicknesses, without common V-preparation and without consumable materials, as well as a significantly lower energy consumption, having a significant impact on cost and time savings during production. The secondary benefits are lower distortion, lower residual stresses, fewer micro-inclusions and improved creep-rupture properties [9–17]. Since the 1960s, the A-TIG process was successfully applied to weld a number of different materials: Titanium alloys, austenitic, ferritic, duplex stainless steels, high strength alloyed and unalloyed steels, nickel-based alloys, etc., as summarised in [18], as well as to weld dissimilar alloys [10,19].

There are several effects that might have a significant influence on the increase in penetration, which is accompanied by the narrowing of the weld. Welds in A-TIG change from wide shallow type (with a relatively low depth-to-width ratio) to a deep, narrow weld (with several times higher depth-to-width ratio) [20,21]. The two major factors responsible for an improved penetration depth appear to be the reversal of Marangoni convection and arc constriction [18]. Marangoni convection is a surface-tension-driven convection depending on the surface tension gradient in the fluid. As fluid flows from areas where surface tension is lower towards areas where it is higher, the reversal of surface tension influences the flow of the molten metal and at the same time can influence the shape of the weld metal. In TIG welding, the flow is from the center of the weld pool towards the fusion boundary, influencing the occurrence of the wider and less deep weld. By reversing the position of these areas, the molten metal can also be reversed, flowing from the fusion boundary towards the center, resulting in a narrower and deeper weld [22,23].

The increase in penetration can be achieved in molten metals containing small amounts of impurities such as sulphur, but obviously the alternative in the form of oxygen might be more attractive [24]. Zou et al. [24] applied a double flow plasma torch with oxygen gas added to the outer flow, to achieve a depth-to-width ratio of up to 0.8. Another approach is to use coatings, based on $SiO_2$, $TiO_2$, $MoO_3$, $Cr_2O_3$, NiO, and CuO powder in a solvent, usually ethanol or acetone. This approach also influences arc constriction effect, which is achieved by the electronegativity of the coating, especially by the presence of Si and Ti [25,26]. Using this approach, even higher depth-to-width bead ratios can be achieved.

Tseng et al. [2] reported a depth-to-width ratio of 1.08 using $SiO_2$ nanoparticle-based coating applied to UNS S31603 stainless steel. In the work by Vora and Badheka [21] on reduced-activation ferritic/martensitic steel, a range of coatings was tested, based on $Al_2O_3$, $Co_3O_4$, CuO, HgO, $MoO_3$, and NiO, of which the most effective were $Co_3O_4$ and CuO, due to the identified reversed Marangoni effect and arc constriction effect. Also, considerable work was done on studying complex coatings, containing different types of powders. Venkatesan et al. [27] studied the effect of three different types of powders, $SiO_2$, $TiO_2$, and $Cr_2O_3$, in different ratios. They found that the mixtures of powders have a more pronounced effect on penetration depth, more specifically, the mixture of $SiO_2$ and $TiO_2$ having the highest effectiveness in welding of AISI 409 ferritic stainless steel. The influence of particle size was also studied [2], where 75 μm and 40 nm $SiO_2$ and 95 μm and 50 nm $Al_2O_3$ micro and nanoparticles were used as key constituents in powders. It was shown that the influence of particle size of $Al_2O_3$ did not have a crucial influence on increase in penetration, unlike $SiO_2$.

In this study, the influence of metallic oxide nano- and submicron particles in different ratios on the performance of A-TIG welding of austenitic stainless steels was analysed. The results were correlated to true particle size results. Also, molten metal flow model was developed based on microstructural analysis. The A-TIG process was compared to TIG with consumable material applied, to evaluate the possibility to avoid the application of consumable wire. Special attention was paid to the problem of critical raw materials (CRMs) for the European Union. Namely, the consumables used in the welding of austenitic stainless steels contain critical raw materials (CRM) or nearly CRMs and relatively expensive materials such as chromium, nickel, and silicon metal [28,29], the use of which can be suppressed by using the proposed approach and it became the motivation for this work.

## 2. Materials and Methods

The base metal used was AISI 304 (X5CrNi18-10) stainless steel in the form of 10-mm thick plates. The chemical composition of this material was <0.03% C, 0.5% Si, 1.3% Mn, <0.008% Si, 18.03% Cr, 0.003% P, 0.01% Al, 0.41% Cu, 9.51% Ni, 0.012% Sn, 0.07% V, and the remaining Fe. The two types of coatings used during the A-TIG welding were based on $TiO_2$ oxides of 20-nm nanoparticles and 0.3-μm submicron particles (Figure 1). As part of this investigation, six different mixtures (by weight percentages) were prepared, containing 5 wt. % of particles in acetone (($CH_3)_2CO$) and were referenced to the control sample without the coating and consumable material (specimen 0).

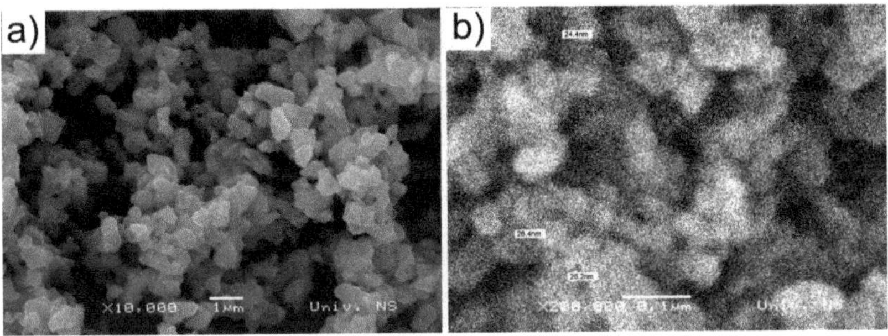

**Figure 1.** Basic components used: (**a**) 0.3-μm Submicron particles used in the coating, (**b**) nanoparticles used in the coating.

(1)   All-submicron particles (5M);
(2)   20% nano and 80% submicron particles (4M1N);
(3)   40% nano and 60% submicron particles (3M2N);
(4)   60% nano and 40% submicron particles (2M3N);
(5)   80% nano and 20% submicron particles (1M4N);
(6)   All-nanoparticles (designated as 5N).

Weighing was done on a Tehtnica Type 2615 analytic balance (Zelezniki, Slovenia), while mixing of the oxide particles into the carrier solvent was done with a Tehtnica MM530 magnetic stirrer (Zelezniki, Slovenia), for 600 s. The size of the particles in the liquid component was determined by the application of a Zetasizer Nano ZS device (Malvern Instruments, Malvern, UK).

The coating was manually applied over the base material with a 10-mm brush, in a layer having a width of approximately 20 mm. The welding-remelting was done on EWM Tetrix 230AC/DC device (Mündersbach-Westerwald, Germany), with 200 A DCEN current and by using a nozzle diameter of 12.7 mm. The gap from the electrode tip to the base metal surface was kept as 2 mm. The process was done with 2.4-mm tungsten electrode containing 2% thorium (with red color mark). To study the electrode tip geometry, three different shapes were used: Conical 90° sharp tip (designated as S), conical 90° with 0.5-mm frustum (F) at the tip, and blunt tip (B). Argon gas flow rate was set at 12 L/min, while the welding speed was maintained at 100 mm/min, along with the center of 50-mm width of the stainless-steel strip. Welding extension was 6 mm. Specimen 0 was prepared by machining a 2-mm-deep square V-groove. This was done to facilitate the application of 0.8-mm coil wire made of AISI 308 austenitic stainless steel with the following nominal chemical composition: ≤0.08% C, ≤2% Mn, ≤0.045% P, ≤0.03% S, ≤1% Si, 19–21% Cr, and 10–12% Ni. Other specimens (1–6) were welded without the consumables.

Post-weld characterisation was done in terms of macro- and microstructure examination, and microhardness. Macro- and microstructure examinations were done by cutting, grinding (abrasive papers), and polishing (diamond suspensions 6–$\frac{1}{4}$ μm), followed by aqua regia etching. Weld width and depths were measured, while depth-to-width ratios were calculated. Also, microstructures in various typical places such as the weld bead (WB), heat-affected zone (HAZ), and base metal (BM) were examined on a Leitz Orthoplan light microscope (Oberkochen, Germany). The light microscope was also used for accurate measurement of WB width and penetration.

Vickers microhardness was done along line 1, 1 mm under the surface; along line 2, 1 mm above the bottom of the WB through BM, HAZ, WB, HAZ, and WB; and finally, through the center of the WB, perpendicular to lines 1 and 2, starting at 1 mm under the surface, to the bottom of the WB, HAZ, and BM, Figure 2. The distance between the indentations was 0.5 mm. Indentation loading was 0.981 N (100 gf) on Wilson Tukon 1102 (Uzwil, Switzerland) device.

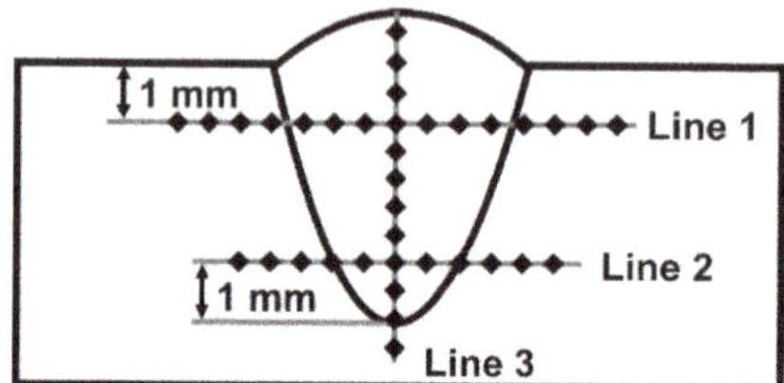


**Figure 2.** Microhardness measurement scheme [30].

## 3. Results

### 3.1. Particle Size Distribution in the Coating

Particle size distribution in the coating is shown in Figure 3. Despite the use of nominally nano- and submicron-sized particles, all specimens showed significant agglomeration. The smallest detected particles were 0.25 µm, while the largest were in the range of 15–16 µm. Particles of size up to 1 µm were seen more commonly in all the specimens. There was a significant difference between the specimen containing only submicron particles (5M) and other specimens, containing also nanoparticles. In specimens 4M1N and 3M2N, the smallest particles detected were of 0.25 µm, and in specimens 2M3N, 1M4N, and 5N the smallest detected particles were of 0.29-µm size, while in all-submicron-particle mixture 0.95-µm particles were the smallest. Although nano-based mixture (5N) showed it to be more effective in the sense of containing smaller-sized particles than the specimen containing submicron-sized particles (5M), the highest amount of the smallest particles were found in mixtures, containing both submicron and nanoparticles (3M2N, 2M3N, and 1M4N).

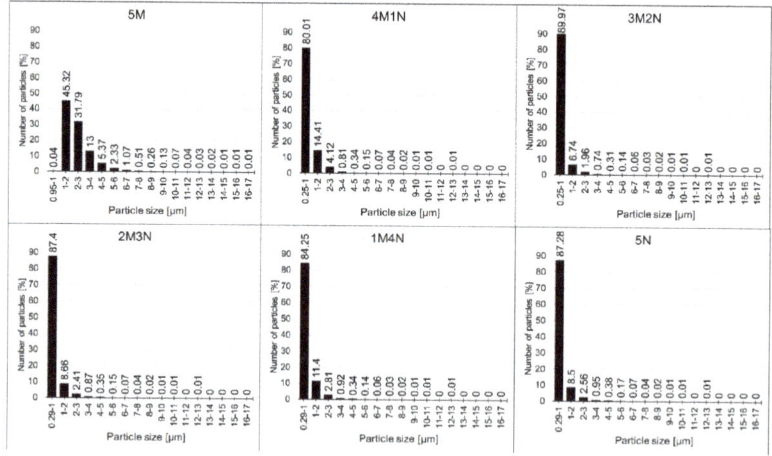

**Figure 3.** Particle size distribution in the solvent.

### 3.2. Macro and Weld Bead Dimensions

Macro images of welds obtained without and with coating, different electrode tip profiles, with indicated depths, widths, and depth-to-width ratios and weld surfaces are shown in Figure 4. It is difficult to establish a direct correlation between the application of coating type and surface areas of weld metals. Several weld metal shape types obtained by A-TIG can be identified. Without the coating (specimens 0), the shape of the weld was flat, semi-elliptical in shape, and could be attributed also to coatings containing submicron particles (specimens 5M) and predominantly submicron particles (specimen 4M1N) regardless of the electrode tip used. Some isolated examples also exist, relating to sharp tip (S) electrode, containing the majority or all nanoparticles in the coating (specimens 1M4N-S and 5N-S). Specimens containing a balanced amount of submicron and nanoparticles in the coating regardless of the electrode tip geometry as well as predominant-nano and all-nanoparticle coating with frustum and blunt tips (designated as F and B) exhibited pronounced change in the shape of the weld, starting from the least pronounced, nearly semicircular in specimen 2M3N-F, to an almost hour-glass shape in 3M2N-F and 1M4N-F. These three specimens obtained with frustum-shaped tip, on the other hand, indicated that relatively small variations in submicron and nanoparticle content may induce large differences both in weld shape and dimensions.

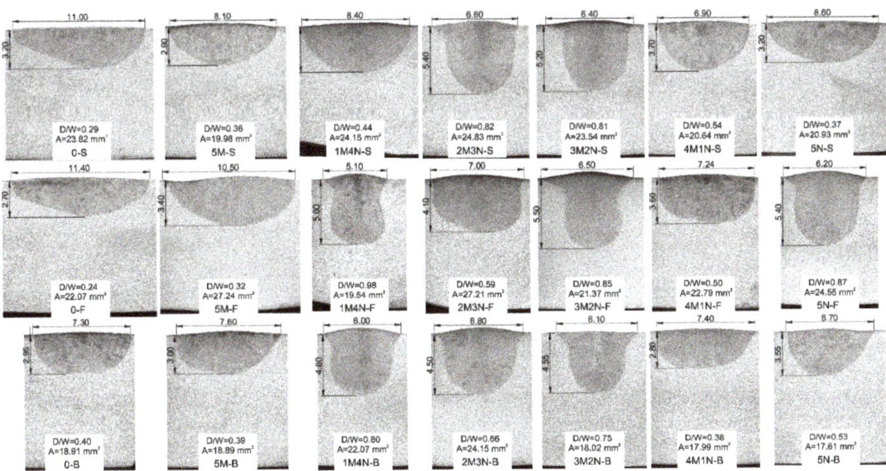

**Figure 4.** Macrostructure of specimens welded without and with coating, with indicated depths, widths, depth-to-width ratios and weld surfaces.

Depths of penetration of A-TIG specimens were higher than those of the control TIG specimens. Also, there were differences between specimens welded with different types of electrodes. Maximum values of penetration depths were higher in specimens welded with S and F electrodes than with the B electrode. Coatings containing only nanoparticles (5N) influenced the lower depth of penetration compared to other mixtures, including submicron particles. The highest penetration was obtained with the combination of submicron and nanoparticles in the mixture, such as 1M4N, 2M3N, and 3M2N, regardless of the electrode tip geometry used. An increase in the penetration depth caused narrowing of the weld width. The same can be established by measurement of the depth-to-width ratios, also given in Figure 3, reaching almost the value of 1 in specimen 4M1N-F.

### 3.3. Microstructure

Microstructures of specimens obtained from light microscope 0-F, 5M-F, 1M4N-F, and 5N-F are presented in Figures 5–7. In Figure 5, weld metal microstructures of 0-F, 1M4N-F, and 5N-F are shown. All microstructures corresponded to the typical dendritic morphology found in weld metal in stainless steels.

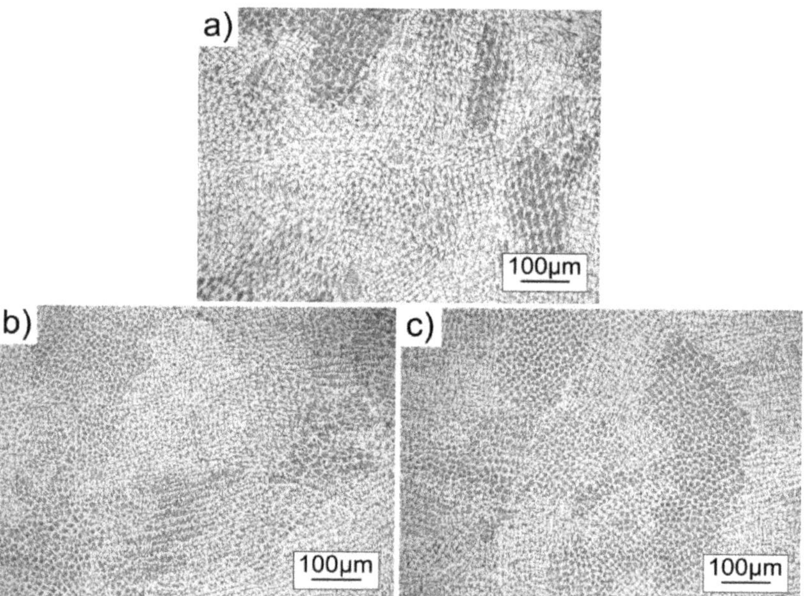

**Figure 5.** Weld metal microstructure of tested specimens: (**a**) 0-F, (**b**) 1M4N-F, (**c**) 5N-F.

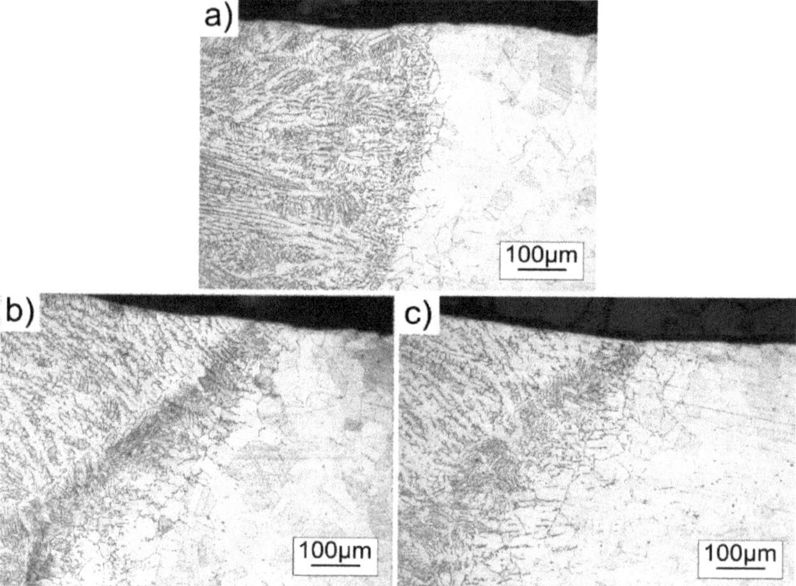

**Figure 6.** Microstructures near fusion line under the specimen surface: (**a**) 0-F, (**b**) 1M4N-F, (**c**) 5N-F.

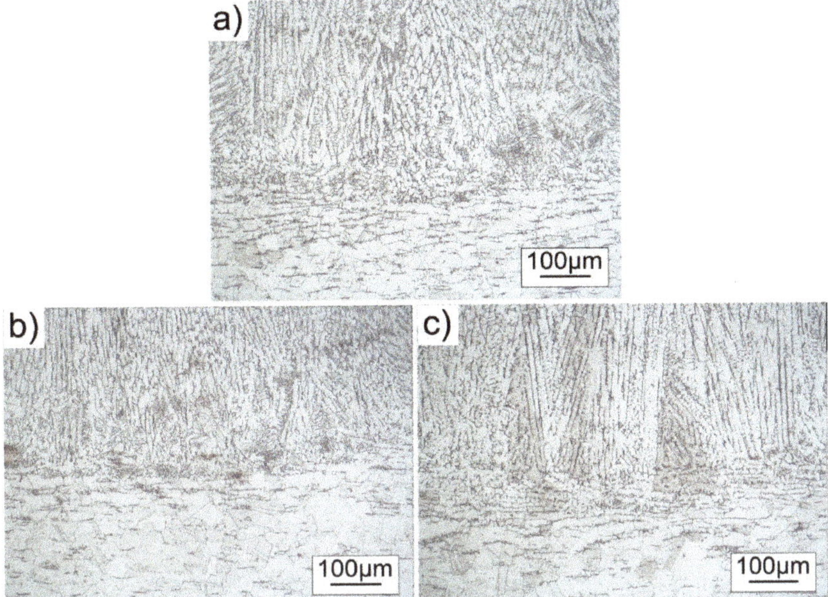

**Figure 7.** Microstructure of the specimens near fusion line at the bottom of the weld: (**a**) 0-F, (**b**) 1M4N-F, (**c**) 5N-F.

In Figure 6, the microstructure near the fusion line at the surface is shown, while in Figure 7, the fusion line at the bottom of the weld is depicted. There was a considerable difference between specimens obtained without the coating (0-F) and specimens obtained with the coating (1M4N-F and 5N-F). In specimen 0-F, austenitic grain coarsening was noticeable near the surface, while in specimens 1M4N-F and 5N-F in the bottom, under the weld line, in the base metal.

*3.4. Vickers Microhardness*

Vickers microhardness testing was done on the same specimens selected for microstructure testing: 0-F, 5M-F, 1M4N-F, and 5N-F. Microhardness values were measured in three lines, as explained in the Experimental part and graphically presented earlier in Figure 2. In Figure 8, values measured along lines 1 and 2, while in Figure 9, values along line 3 are shown. In specimen 0-F, minimal values occur near the fusion line (hollow marks, indicated by arrows), at the position just under the specimen surface (measurement line 1). On the other hand, in specimens 5M-F, 1M4N-F, and 5N-F, minimal values are obtained under the weld (hollow marks, indicated by arrows). These values closely corresponded to the occurrence of coarsened austenitic grains shown in the preceding section.

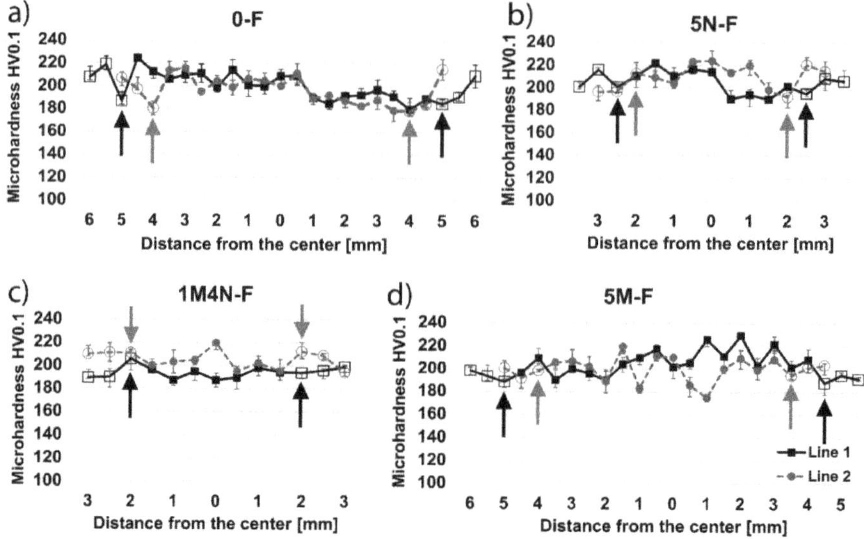

**Figure 8.** Microhardness distribution for specimens welded with frustum-tipped electrode, along lines 1 and 2: (**a**) 0-F, (**b**) 5M-F, (**c**) 1M4N-F, (**d**) 5N-F.

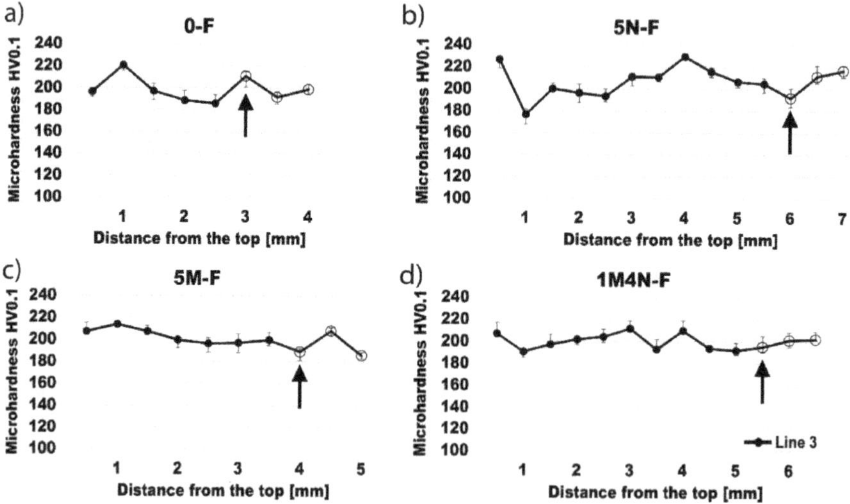

**Figure 9.** Microhardness distribution for specimens welded with frustum-tipped electrode, along line 3: (**a**) 0-F, (**b**) 5N-F, (**c**) 5M-F, (**d**) 1M4N-F.

## 4. Discussion

In this paper, different $TiO_2$ submicron and nanoparticle ratios were interspersed to produce A-TIG coatings, used for welding by 2% thoriated tungsten electrodes with various electrode tip profiles. The existence of coating did not influence the weld metal surface area. This was in contrast to the work of Tseng and Lin [2], who obtained significantly increased weld metal surface areas with the coating and larger surface areas with the coating based on $SiO_2$ and $Al_2O_3$ nanoparticles compared to the ones using microparticles.

The highest depth of penetration was obtained with frustum-type electrode, in specimen 3M2N-F, closely followed by the specimen 5N-F and 2M3N-S, obtained with frustum and sharp electrodes, respectively. $TiO_2$ nanoparticles proved to be more effective than submicron particles, but the mix of the two proved to be the most effective. A similar trend was noticed in D/W ratios. Nanoparticles proved to be more effective than submicron particles in the coating, especially with frustum electrodes used. This result is in agreement with the results reported by Tseng and Lin [2], who demonstrated that the $SiO_2$ nanoparticles were more effective in achieving an increase in penetration depth than the microparticles of the same type. However, in [2], no significant gain in penetration was achieved by using $Al_2O_3$ nanoparticles versus microparticles. The advantage of nanoparticles versus larger submicron- or micron-sized particles could be attributed to a higher effectiveness of smaller coating particles. In arc heating, the thermal dissociation and decomposition of smaller particles occurred much more readily than in larger particles, due to their higher specific surface area. However, the nominal size of the particles used for the coating did not reflect their performance ideally, due to agglomeration. Therefore, a much more accurate indicator of particle effectiveness can be obtained by Zeta sizer true particle size results.

Blunt electrodes generally proved the least effective. The results showed an inferior performance in terms of weld depth which may be explained by the width of the electric arc and the corresponding width of the coating that is heated, evaporated, and thermally dissociated. Blunt electrodes offer a relatively narrow and deep weld when the welding is done without the coating, Figure 4. This is in agreement with other reports [31,32], where blunt electrodes offered higher penetration, versus sharpened electrode, which was reported to offer wider and shallower welds. The main reason is a wider electric arc, with the energy spread over a larger area. However, in A-TIG, with coating applied, a wider electric arc also influences heating and vaporization of the coating applied to the surface. That means a blunt electrode, in spite of the fact that it theoretically offers the highest penetration (in conventional TIG), when used with the coating, its effect on penetration was inferior to that of the conical and frustum electrodes. The main reason in obtaining a lower penetration compared to other electrode geometries lies in a lower width covered by the arc (Figure 10), and subsequently, a lower amount of oxides that were vaporized, dissociated, and decomposed.

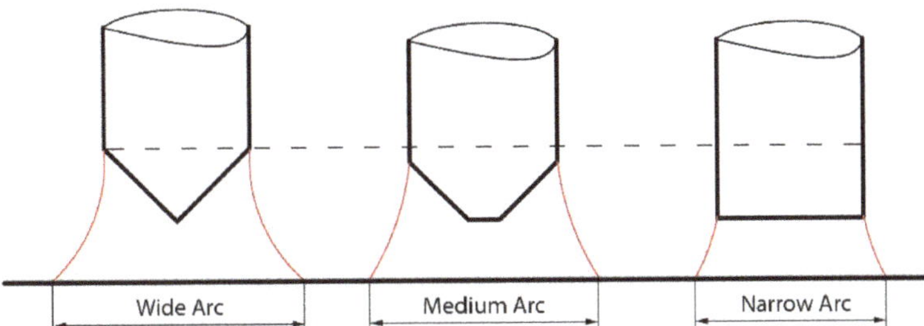


**Figure 10.** Arc width in relation to electrode geometry: Sharp electrode offers the widest arc, followed by frustum and blunt electrode.

Frustum-type electrode tip results in a combined concentrated-spread arc, offering a combination of a higher penetration due to the flattened tip and the vaporization, dissociation, and decomposition of the oxides. This is particularly obvious in the penetration depth of the specimen containing nano particles, 5N-F.

Of all tested coatings, the most effective were mixtures of nano and submicron $TiO_2$ particles (specimens 3M2N, 2M3N, 1M4N), containing between 40% and 80% of nanoparticles and 20–60% of submicron particles. This may be due to the existence of submicron and nanoparticle agglomerates. Agglomerates have a relatively low cohesive strength, due to the presence of relatively weak secondary bonds between particles, usually Van der Waals forces, hydrogen or capillary [33–35]. By mixing nano and submicron particles, their agglomerates suffer random collisions, as reported by Dongguang et al. [36], causing a grinding effect, leading to the obtaining of smaller particles in the coating mixture, as revealed by Zeta sizer, Figure 3. A stochastic nature of these collisions may influence variable performance of the coatings used for welding specimens 2M3N and 1M4N with different types of electrodes. Also, a variable performance of all-nano coatings (5N) can also be the result of variable agglomeration between nanoparticles.

The material flow model proposed in accordance with the results obtained in this work is summarised in Figure 11. In specimens obtained without the coating, the hardness near the fusion line at the surface had a marked drop compared to the specimens obtained with the coating applied. This was in good agreement with microstructures in these zones, with a decreased hardness closely corresponding to the coarser austenite grains in respective zones. This indicates that the hot fluid flows from the surface (in specimens obtained without the coating), towards the fusion boundary of the melt pool, transferring the heat into the base metal. This heat caused the austenite grains to grow, resulting in coarser grains near the fusion line, under the surface. The austenite grains under the weld remained unchanged. On the other hand, a reversed Marangoni convection in specimens obtained with the coating caused a hot melt to flow inwards and push into the base material. This caused a higher penetration, but, as a side effect, a heat transfer towards the area under the weld metal, transferring heat to this area. As a result, coarsened austenitic grains occurred under the weld. In contrast, austenitic grains near the fusion line just under the surface remained unchanged, since the melt reaching this area already transferred part of its heat. These results are similar to ones presented in [30,37], where $SiO_2$ and $TiO_2$ nanoparticles were used as a basis for the coating. In A-TIG welding, where full penetration is achieved, grain coarsening under the weld metal is not possible, since in case of full penetration, there is no base metal. However, in such an arrangement, it would be necessary to use backing plates to prevent or limit over-penetration.

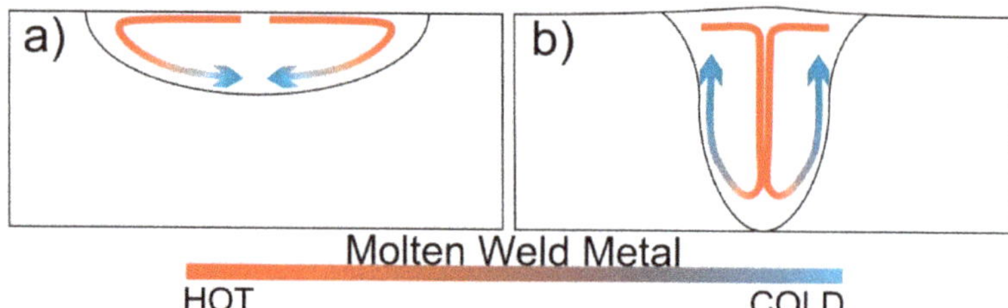


**Figure 11.** TIG and A-TIG metal flows: (**a**) Without the coating, (**b**) with the coating.

Therefore, microstructures and microhardnesses indicated that the reversal of Marangoni convection theory had a significant impact on an increased penetration, D/W ratio, and generally the shape of the weld, as suggested in [18,26,38,39].

## 5. Conclusions

According to the results presented in this work, the following conclusions can be drawn:

1.  The correlation between the coating composition and weld metal surface areas could not be determined, although weld metal areas were larger with frustum and conical tips compared to blunt tip.
2.  Nanoparticles were more effective than submicron particles in increasing the penetration, but a mixture of nano- and micron-sized particles helped achieve the best weld. The main reason of this appears to be the collisions that occurred between agglomerates and submicron particles, resulting in a lower size of particles in the lowest range of sizes that were most effective in increasing the weld penetration.

3. Frustum and sharp electrodes proved to be more consistent in producing high penetration compared to blunt electrodes, due to a narrower arc that affected a narrower width of the coating applied to the base metal surface.

4. Specimens welded without the coating showed an increased grain size near the fusion line in the base metal under the surface, resulting in a decreased hardness in this zone. Contrarily, specimens welded with the coating showed an increased grain size near the fusion line in the base metal under the weld metal, resulting in a decreased hardness in this zone.

5. The main cause of a reduced hardness and increased grain size under the surface and under the weld metal may be attributed to high-temperature material flowing near these zones and heat transfer towards base metal.

6. Material flow direction in the weld pool was the result of Marangoni convection in TIG and reversed Marangoni convection in A-TIG. In TIG, the flow was towards the fusion boundary and along the fusion line, while in A-TIG, the flow towards the center of the weld and to the bottom of the weld was more pronounced resulting in an increase in the weld penetration depth.

**Author Contributions:** S.B. designed the experiment and wrote the paper; M.D. and P.J. performed the experiments; I.Z. provided the resources (devices, materials); B.P. interpreted the data, S.G. and M.S. reviewed and edited the manuscript.

**Acknowledgments:** This article is based on the work from COST Action "Solutions for Critical Raw Materials under Extreme Conditions", supported by COST Action 15102 under the auspices of H2020. Saurav Goel sincerely acknowledges the financial support obtained from various funders including the RCUK (Grant No. EP/S013652/1 and EP/S036180/1), H2020 (EURAMET EMPIR A185 (2018)), Royal Academy of Engineering (Grant No. IAPP18-19\295), and Newton Fellowship award from the Royal Society (NIF\R1\191571) that has sustained his research laboratory and its group members.

**Conflicts of Interest:** The authors declare no conflict of interest.

## References


1. Tseng, K.-H.; Hsu, C.-Y. Performance of activated TIG process in austenitic stainless steel welds. *J. Mater. Process. Technol.* **2011**, *211*, 503–512. [CrossRef]

2. Tseng, K.-H.; Lin, P.-Y. UNS S31603 Stainless Steel Tungsten Inert Gas Welds Made with Microparticle and Nanoparticle Oxides. *Materials* **2014**, *7*, 4755–4772. [CrossRef] [PubMed]

3. Gurevich, S.M.; Zamkov, V.N.; Kushnirenko, N.A. Improving the penetration of titanium alloys when they are welded by argon tungsten arc process. *Avtom. Svarka.* **1965**, *18*, 1–5.

4. Huang, H.Y. Argon–hydrogen shielding gas mixtures for activating coating-assisted gas tungsten arc welding. *Metall. Mater. Trans. A* **2010**, *41*, 2829–2835. [CrossRef]

5. Huang, H.Y. Effects of shielding gas composition and activating coating on GTAW weldments. *Mater. Des.* **2009**, *30*, 2404–2409. [CrossRef]

6. Paskell, T.D. Gas Tungsten Arc Welding Coating. U.S. Patent 5,804,792A, 9 August 1998.

7. Tseng, K.H.; Wang, N.S. Welding Activated Coating for Structural Alloy Steels. U.S. Patent 20,160,167,178A1, 16 June 2016.

8. Muthukumaran, V.; Bhaduri, A.K.; Raj, B. Penetration Enhancing Coating Formulation for Tungsten Inert Gas (TIG) Welding of Austenitic Stainless Steel and Its Application. U.S. Patent 8,097,826, 17 January 2012.

9. Dhandha, K.H.; Badheka, V.J. Effect of activating coatings on weld bead morphology of P91 steel bead-on-plate welds by coating assisted tungsten inert gas welding process. *Mater. Manuf. Process.* **2015**, *17*, 48–57. [CrossRef]

10. Nayee, S.G.; Badheka, V.J. Effect of oxide-based coatings on mechanical and metallur-gical properties of Dissimilar activting coating assisted-tungsten inert gas welds. *Mater. Manuf. Process.* **2014**, *16*, 137–143. [CrossRef]

11. Modenesi, P.J.; Apolinário, E.R.; Pereira, I.M. TIG welding with single-component coatings. *J. Mater. Process. Technol.* **2000**, *99*, 260–265. [CrossRef]

12. Lin, H.L.; Wu, T.M. Effects of activating coating on weld bead geometry of Inconel718 alloy TIG welds. *Mater. Manuf. Process.* **2012**, *27*, 1457–1461. [CrossRef]

13. Kou, S. *Welding Metallurgy*, 2nd ed.; Wiley—Interscience Publishers: Hoboken, NJ, USA, 2003; pp. 116–117.

14. Vasudevan, M.; Bhaduri, A.K.; Raj, B.; Prasad, R.K. Genetic algorithm based com-putational model for optimizing the process parameters in A-TIG welding of 304LN and 316LN stainless steels. *Mater. Manuf. Process.* **2007**, *22*, 641–649. [CrossRef]

15. Chandrasekhar, N.; Vasudevan, M. Intelligent modeling for optimization of A-TIG welding process. *Mater. Manuf. Process.* **2010**, *25*, 1341–1550. [CrossRef]

16. Maduraimuthu, V.; Vasudevan, M.; Muthupandi, V.; Bhaduri, A.K.; Jayakumar, T. Study of the effect of activated coating on the microstructure and mechanical properties of mod. 9Cr-1Mo steel. *Metall. Mater. Trans.* **2012**, *43*, 123–132. [CrossRef]

17. Sakthivel, T.; Vasudevan, M.; Laha, K.; Parameswaran, P.; Chandravathi, K.S.; Paneer-Selvi, S.; Maduraimuthu, V.; Mathew, M.D. Creep-rupture behavior of 9Cr-1.8W-0.5Mo-VNb P92 ferritic steel weld joint. *Mater. Sci. Eng.* **2014**, *591*, 111–120. [CrossRef]

18. Vidyarthy, R.S.; Dwivedi, D.K. Activating coating tungsten inert gas welding for enhanced weld penetration. *J. Manuf. Process.* **2016**, *22*, 211–228. [CrossRef]

19. Kuo, C.H.; Tseng, K.H.; Chou, C.P. Effect of activated TIG coating on performance of dissimilar welds between mild steel and stainless steel. *Key Eng. Mater.* **2011**, *479*, 74–80. [CrossRef]

20. Vora, J.J.; Badheka, V.J. Improved Penetration with the Use of Oxide Coatings in Activated TIG Welding of Low Activation Ferritic/Martensitic Steel. *Trans. Indian Inst. Met.* **2016**, *69*, 1755–1764. [CrossRef]

21. Vora, J.J.; Badheka, V.J. Experimental investigation on mechanism and weld morphology of activated TIG welded bead-on-plate weldments of reduced activation ferritic/martensitic steel using oxide coatings. *J. Manuf. Process.* **2015**, *20*, 224–233. [CrossRef]

22. Takeuchi, Y.; Takagi, R.; Shinoda, T. Effect of bismuth on weld joint penetration in austenitic stainless steel. *Weld. Res. Suppl.* **1992**, *71*, 283–290.

23. Mills, K.C.; Keene, B.J.; Brooks, R.F.; Shirali, A. Marangoni effects in welding. *Philos. Trans. R. Soc. Lond. Ser. A* **1998**, *356*, 911–925. [CrossRef]

24. Zou, Y.; Ueji, R.; Fujii, H. Effect of oxygen on weld shape and crystallographic orientation of duplex stainless steel weld using advanced A-TIG (AA-TIG) welding method. *Mater. Charact.* **2014**, *91*, 42–49. [CrossRef]

25. Skvortsov, E.A. Role of electronegative elements in contraction of thearc discharge. *Weld. Int.* **1998**, *12*, 471–475. [CrossRef]

26. Tanaka, M.; Shimizu, T.; Terasaki, T.; Ushio, M.; Koshiishi, F.; Yang, C.-L. Effects of activating coating on arc phenomena in gas tungstenarc welding. *Sci. Technol. Weld Join.* **2000**, *5*, 397–402. [CrossRef]

27. Venkatesan, G.; Goeuge, J.; Sowmyasri, M.; Muthapandi, V. Effet of ternarz coatings on depth of penetration in A-TIG welding of AISI 409 ferritic stainless steel. *Proc. Mater. Sci.* **2014**, *5*, 2402–2410. [CrossRef]

28. Report on Critical Raw Materials for EU, Report of the Ad-Hoc Working Group on Defining Critical Raw Materials for EU. May 2014. Available online: http://mima.geus.dk/report-on-critical-raw-materialsen.pdf (accessed on 28 August 2019).

29. Luisa Grilli, M.; Bellezze, T.; Gamsjäger, E.; Rinaldi, A.; Novak, P.; Balos, S.; Piticescu, R.; Letizia Ruello, M. Solutions for Critical Raw Materials under Extreme Conditions: A Review. *Materials* **2017**, *10*, 285. [CrossRef] [PubMed]

30. Balos, S.; Dramicanin, M.; Janjatovic, P.; Zabunov, I.; Klobcar, D.; Busic, M.; Luisa Grilli, M. Metal oxide nanoparticle-based coating as a catalyzer for A-TIG welding: Critical raw material perspective. *Metals* **2019**, *9*, 567. [CrossRef]

31. Key, J.F. Anode/Cathode Geometry and Shielding Gas Interrelationships in GTAW Electrode tip geometry and groove geometry must be compatible to ensure arc stability. In Proceedings of the 61st AWS Annual Meeting, Los Angeles, CA, USA, 13–18 April 1980; pp. 364–370.

32. Mannion, B.; Heizman, J., III. Setting up and Determining Parameters for Orbital Tube Welding. Available online: http://www.pro-fusiononline.com/feedback/fab-may99.htm (accessed on 15 June 2019).

33. Balos, S.; Pilic, B.; Markovic, D.; Pavlicevic, J.; Luzanin, O. Poly(Methyl-Methacrylate) Nanocomposites with Low Silica Addition. *J. Prosthet. Dent.* **2014**, *111*, 327–334. [CrossRef] [PubMed]

34. Balos, S.; Pilic, B.; Petrovic, D.; Petronijevic, B.; Sarcev, I. Flexural strength and modulus of autopolimerized poly(methyl methacrylate) with nanosilica. *Vojnosaniteski Pregled.* **2018**, *75*, 564–569. [CrossRef]

35. Elshereksi, N.W.; Mohamed, S.H.; Arifin, A.; Mohd Ishak, Z.A. Effect of filler incorporation on the fracture toughness properties of denture base poly(methyl methacrylate). *J. Phys. Sci.* **2009**, *20*, 1–12.

36. Dongguang, W.; Rajesh, D.; Robert, P. Mixing and characterization of nanosized powders: An assessment of different techniques. *J. Nanopart. Res.* **2002**, *4*, 21–41.
37. Dramicanin, M.; Balos, S.; Janjatovic, P.; Zabunov, I.; Grabulov, V. Activated Flux TIG Welding of Stainless Steel Pipes. *Chem. Ind. Chem. Eng. Q.* **2019**. [CrossRef]
38. Roper, J.R.; Olson, D.L. Capillarity effects in the GTA weld penetration of 21-6-9 stainless steel. *Weld J.* **1978**, *57*, 103s–107s.
39. Dong, C.; Zhu, Y.; Chai, G. Preliminary study on the mechanism of arc welding with the activating coating. *Aeronaut. Manuf. Technol. Suppl.* **2004**, *6*, 271–278.

*Article*

# Metal Oxide Nanoparticle-Based Coating as a Catalyzer for A-TIG Welding: Critical Raw Material Perspective

Sebastian Balos [1,*], Miroslav Dramicanin [1], Petar Janjatovic [1], Ivan Zabunov [2], Damjan Klobcar [3], Matija Busic [4] and Maria Luisa Grilli [5]

[1]   Faculty of Technical Sciences, University of Novi Sad, Trg Dositeja Obradovica 6, 21000 Novi Sad, Serbia; dramicanin@uns.ac.rs (M.D.); janjatovic@uns.ac.rs (P.J.)
[2]   Faculty of Special Technology, Alexander Dubček University of Trenčín, Študentská 2 911 50 Trenčín, Slovakia; ivan.zabunov@tnuni.sk
[3]   Faculty of Mechanical Engineering, University of Ljubljana, Aškerčeva c. 6, 1000 Ljubljana, Slovenia; Damjan.Klobcar@fs.uni-lj.si
[4]   Faculty of Mechanical Engineering and Naval Architecture, University of Zagreb, Ivana Lučića 5, 10002 Zagreb, Croatia; Matija.Busic@fsb.hr
[5]   Italian National Agency for New Technologies, Energy and Sustainable Economic Development (ENEA), Casaccia Research Centre, Via Anguillarese 301, 00123 Rome, Italy; marialuisa.grilli@enea.it
*   Correspondence: sebab@uns.ac.rs; Tel.: +381-21-485-2339

Received: 6 March 2019; Accepted: 19 March 2019; Published: 15 May 2019

**Abstract:** Besides a wide application in corrosion protection, wear resistance increase, providing thermal properties and power conversion, oxide coatings have found an alternative application in welding technology as catalysts of the tungsten inert gas (TIG) welding process. In this paper, the novel approach of fabricating a coating containing nanoparticles based on nanosized $SiO_2$ and $TiO_2$ and their mixtures was applied to the austenitic stainless-steel base metal. It was found that coatings increased depths of penetration, enabling a consumable-free welding. Using this method, the use of several critical and near-critical raw materials (e.g., Si and Cr), as well as the relatively expensive Ni can be completely avoided. The most effective coating in terms of weld penetration consisted of a mixture of nanoparticles, rather than unary oxide coatings based on nanoparticles. A model for liquid weld metal flow is proposed based on the metallographic examination of recrystallized grains and microhardnesses measured near the weld metal, supporting the reversed Marangoni convection theory.

**Keywords:** oxide coating; nanoparticles; TIG welding; penetration depth

## 1. Introduction

The traditional form of oxide coatings has several purposes, mainly to protect the base material against corrosion [1–3] and wear [4–6], to obtain certain thermal properties [7,8], and to be used for power conversion [9,10]. However, a special type of metal-oxide coating can be used in welding technology with a high potential for application in the industrial sector.

One of the most widely used welding processes is gas tungsten arc welding (GTAW) or alternatively called tungsten inert gas (TIG). This process is well known for producing high-quality welds in different metals and alloys, most commonly for different types of stainless steels and non-ferrous alloys based on aluminum, copper, nickel, titanium, etc. [11–13]. Compared to the metal inert gas (MIG), which is used for welding a similar array of materials, TIG suffers from a lower productivity due to a lower welding speed and relatively low penetration, even in the case of currents when an excess of 300 A is used [14]. The low productivity of TIG can be addressed by the introduction of a coating applied before welding over the surface to be welded to act as a catalytic agent of the welding process kinetics. This process is named activated tungsten inert gas (A-TIG). The first of such coatings was developed by Gurevich et al. during the 1960s at the Paton Welding Institute in Kiev, former USSR (now Ukraine). The coating—sprayed or applied by a brush—contains powdered metal oxides such as $TiO_2$, $Al_2O_3$, $SiO_2$, $NiO$, $MnO_3$, $Cr_2O_3$, $CeO_2$, and $V_2O_5$ [15–19], in ethanol, methanol, or acetone solvents [20]. The coated surface is usually approximately 0.15 mm thick [21]. As the highly volatile solvent is quickly dried, the welding will be started shortly after the application to the surface of the base metal. The welding arc can be started once the coating is dried, with the ultimate result of dramatically increased penetration depth ($D$) and narrowed weld width ($W$), achieving a $D/W$ ratio that is close to 0.8 [22], compared to 0.2–0.3 without the coating [16,22]. Usually, nano-oxide particle-based coatings are more effective in achieving both higher penetration and $D/W$ ratios, reaching values over 1 [17]. The most probable cause is a higher specific surface area of nanoparticles compared to the more conventional microparticles. One key mechanism is thought to be the reversal of Marangoni convection. Marangoni convection is driven by weld pool surface tension: fluids flow from areas where the surface tension is lower towards areas where it is higher. In conventional TIG welding, the flow is from the center of the weld pool towards the edges of the weld pool, influencing the occurrence of a wide and shallow weld. By reversing the molten metal flow in A-TIG, flow from the weld pool to the center of the weld occurs, resulting in a significantly narrower and deeper weld [23,24]. Another effect could be the result of the electronegativity of the metallic constituents of oxides. Therefore, Al-, Si-, and Ti-oxides have attracted significant interest in scientific circles [25–27].

There are multiple benefits of using oxide coatings in the A-TIG welding process: The time- and resource-consuming cutting of a V-groove (V-preparation, single V preparation) of the base metal can be avoided, and a simpler and cheaper closed square joint can be used (I-preparation) instead; instead of multi-pass welding, single-pass welding can be sufficient, resulting in significant savings of shielding gas, and there is no need to apply consumable materials in the form of wire rods or wires, therefore allowing a cost reduction of the welding process [20,28]. A special benefit is to avoid the consumption of welding filler material containing critical raw materials for the European Union (EU) [29]. In the case of austenitic stainless steels, some regularly appearing elements such as silicon metal are currently critical for the EU [29]. There are also some elements that were on the previous list of critical raw materials (CRMs) [30]. One example is chromium, which appears with the significant content of 18% and is no longer critical in accordance with the EU's list of critical raw materials [29], but is near-critical and may become critical again in the near future. Another example is nickel, which is important because of its relatively high cost.

In this work, the effects of unary Si and Ti nano-oxide coatings as well as binary coatings containing various ratios of Si and Ti nano-oxides were studied in view of the weld depth, width and $D/W$ ratio, the phenomenology of the process, and the impact on CRMs and near-CRMs, to obtain multiple excellent properties or benefits for the TIG welding process.

## 2. Materials and Methods

In this study, welding was done on AISI 304 austenitic stainless-steel base metal with the following chemical composition: <0.03% C, 0.5% Si, 1.3% Mn, <0.008% S, 18.03% Cr, 0.003% P, 0.010% Al, 0.41% Cu, 9.51% Ni, 0.012% Sn and 0.07% V, with Fe making up the balance. The plates used for welding-re-melting were 10-mm-thick, were water-jet cut to 50 mm width and obtained from a single stainless-steel plate used as provided from the manufacturer, without any heat treatment.

Overall, six different coatings were prepared using 20 nm $TiO_2$ and $SiO_2$ nanoparticles. Nanoparticles at 5 wt. % were mixed with acetone $((CH_3)_2CO)$, having a pH of 7, to produce a uniform mixture. Nanoparticles were weighed using a Tehtnica Type 2615 analytic balance (Zelezniki, Slovenia) and subsequently mixed with the solvent using a Tehtnica mm530 magnetic stirrer (Zelezniki,

Slovenia) for 10 min. To determine the size of the particles in the liquid component, a Zetasizer Nano ZS analyzer (Malvern Instruments, Malvern, UK) was used.

The coating was manually applied to the base material using a brush, coating in layers with a width of approximately 20 mm. The welded specimens, welded with various mixtures, were designated as: 5Si (100% $SiO_2$), 4Si1Ti (80% $SiO_2$; 20% $TiO_2$), 3Si2Ti (60% $SiO_2$; 40% $TiO_2$), 2.5Si2.5Ti (50% $SiO_2$; 50% $TiO_2$), 2Si3Ti (40% $SiO_2$; 60% $TiO_2$), 1Si4Ti (20% $SiO_2$; 80% $TiO_2$), and 5Ti (100% $TiO_2$). Specimen 0 was welded without the coating for the purpose of comparison and evaluation of the effects of coatings.

Welding was done using an EWM Tetrix 230 AC/DC welding power source (Mündersbach-Westerwald, Germany) with 200 A DCEN current and by the application of a 12.7 mm diameter nozzle. The electrode tip was sharpened to 90°. Tungsten with a 2% thorium electrode having a 2.4 mm diameter was used. The distance between the electrode tip and the base material was 2 mm, while the electrode was 6 mm out of the torch. Welding speed was 100 mm/min for all specimens. In specimen 0, the base metal was prepared using a 2-mm-deep V-groove with an angle of 90°. Consumable material in the form of a 0.8-mm coil wire made of AISI 308 (≤0.08% C; ≤2%; ≤0.045% P; ≤0.03% S; ≤1% Si; 19%–21% Cr; 10%–12% Ni) was continuously added to the weld pool. Specimens welded with a coating were welded without the V-groove and consumable material.

The characterization of welds in order to evaluate the effectiveness of produced coatings comprised the following techniques: macro testing, microstructure examination, and microhardness, done in various weld areas (weld material, heat-affected zone, and base material). Macro and microstructural examination was done after a common metallographic preparation procedure consisting of the following: cutting, mounting into a 20 mm diameter polyethylene cup, grinding (using 150–2000 grit abrasive paper) and polishing (using 6, 3, 1, and 0.5 μm diamond suspension). Etching was done using aqua regia. Metallographic specimens were examined using a Leitz Orthoplan light microscope (Oberkochen, Germany), which was also used to measure the crucial weld dimensions (width—*W* and depth—*D*). Microhardness was measured using a Wilson Tukon 1102 (Uzwil, Switzerland) Vickers measurement device with a 1 kg load. Vickers microhardness measurement was done in parallel to the surface, 1 mm under it and 1 mm from the bottom of the weld metal. Microhardness was also measured through the weld metal, from the weld face to the bottom of the weld metal. The schematic depiction of the microhardness measurement is shown in Scheme 1.

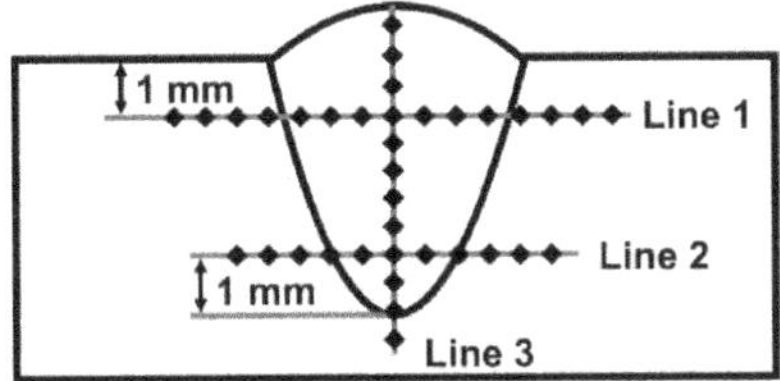


**Scheme 1.** Microhardness measurement scheme.

## 3. Results

### 3.1. Particle Size Fractions

Particle size fractions are shown in Figure 1. In Figure 1, only fractions of up to 4 μm are shown, balance values of up to 17 μm were omitted to enable compactness of the chart. The largest number of particles were up to 1 μm, while larger fractions were progressively smaller as the particle size range was higher. This occurred in spite of the nominal nanoparticles that were used, indicating that intensive agglomeration was present. The fraction of particles having a size up to 1 μm was higher in all $TiO_2$ compared to all $SiO_2$ nanoparticle coating solvents, but the highest number of detected particles in this range was in coatings based on mixtures of $SiO_2$ and $TiO_2$ particles. The maximum particle size fraction was reached in specimen 2Si3Ti.

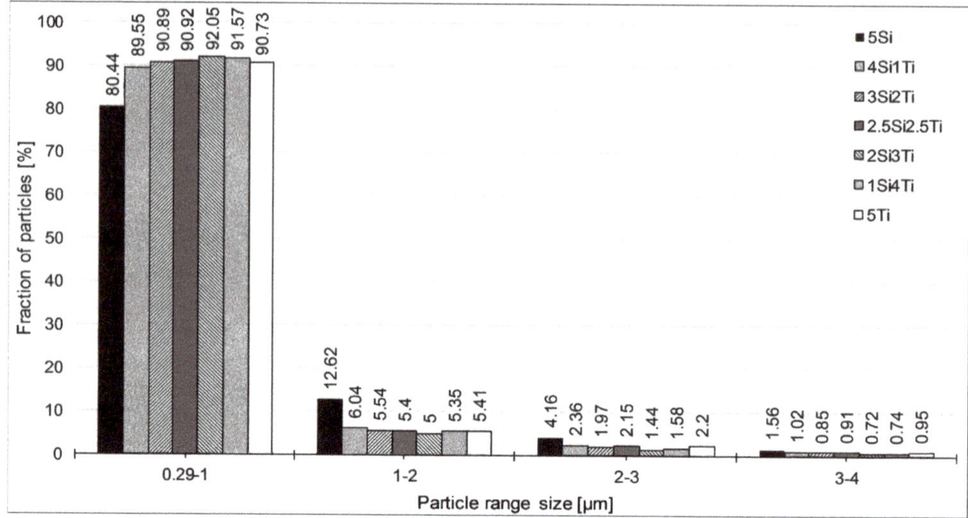

**Figure 1.** Zetasizer results.

*3.2. Macro and Weld Bead Dimensions*

Macro images of obtained weld beads are presented in Figure 2, along with the width (*W*), depth (*D*), depth-to-width ratios (*D/W*), and specimen designation. Weld bead measurements were averages of measurements taken from five macro images reported along with standard deviations. In Figure 2, both the conventionally welded–re-melted specimen (0) and specimens welded with the coating are shown. It can be seen that the introduction of the coating significantly influenced the melted profile of the weld. Firstly, the weld bead width decreased. It was roughly half of that of specimen 0's weld bead width. More importantly, weld depth was significantly increased. Specimen 5Si obtained with a $SiO_2$ nanoparticle coating had a lower penetration depth than 5Ti, obtained with a coating containing $TiO_2$ nanoparticles. The highest penetration was obtained with specimen 2Si3Ti. Deviations reported in Figure 2 were on the same order of magnitude for all specimens. There was no positive correlation between the increase in penetration and the narrowing of the weld, as for the depth-to-width ratios (*D/W*). However, *D/W* ratios increased from 0.28 in specimen 0 to 0.66 (specimen 2Si3Ti) and 0.79 (specimen 1Si4Ti) on average.

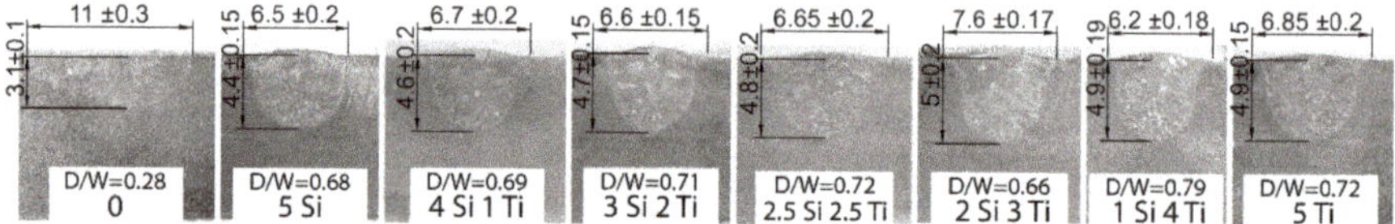

**Figure 2.** Macro images of specimens welded without (specimen 0) and with coating (specimens 5Si, 4Si1Ti, 3Si2Ti, 2.5Si2.5Ti, 2Si3Ti, 1Si4Ti, and 5Ti), with the indicated weld bead widths, penetration depth/width ratios (*D/W*), and standard deviations.

*3.3. Microstructures*

Microstructures identified within and near the melting line are shown in Figures 3–5. The analysis will mainly refer to the comparison between specimen 0, obtained without the coating and with consumable material, and specimens obtained with the coating and without consumable material. Weld metal microstructures of specimens 0, 4Si1Ti, 3Si2Ti, and 5Ti are shown in Figure 3. It can be seen that the microstructure was typically dendritic, a common feature for all the tested specimens.

Some differences appeared to exist, but they were the result of a different orientation of dendrites and various cross sections that were obtained.

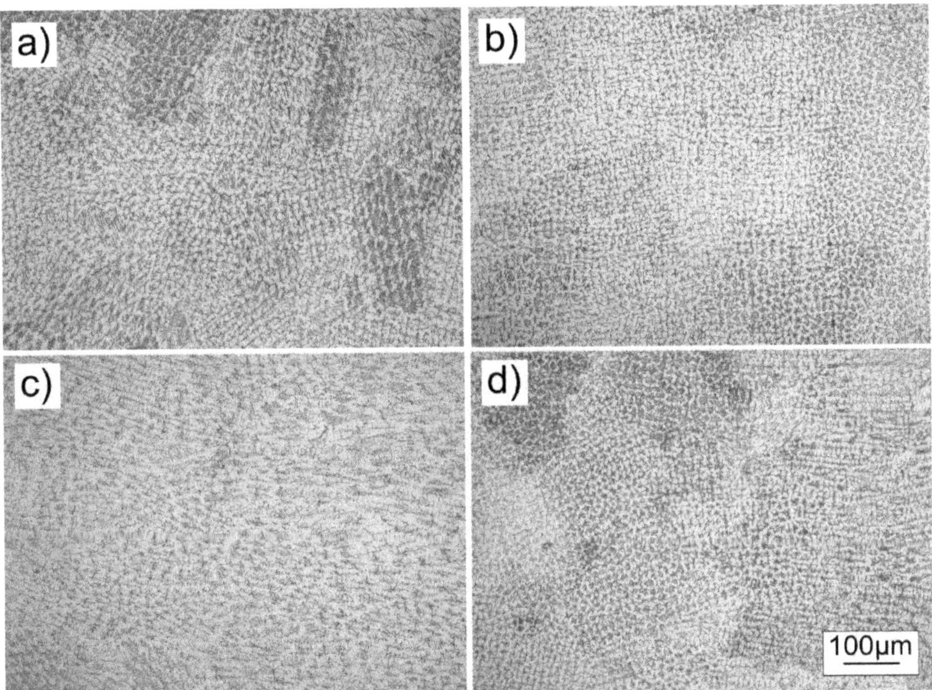

**Figure 3.** Microstructures of weld metals of specimens: (**a**) 0; (**b**) 4Si1Ti; (**c**) 3Si2Ti; (**d**) 5Ti.

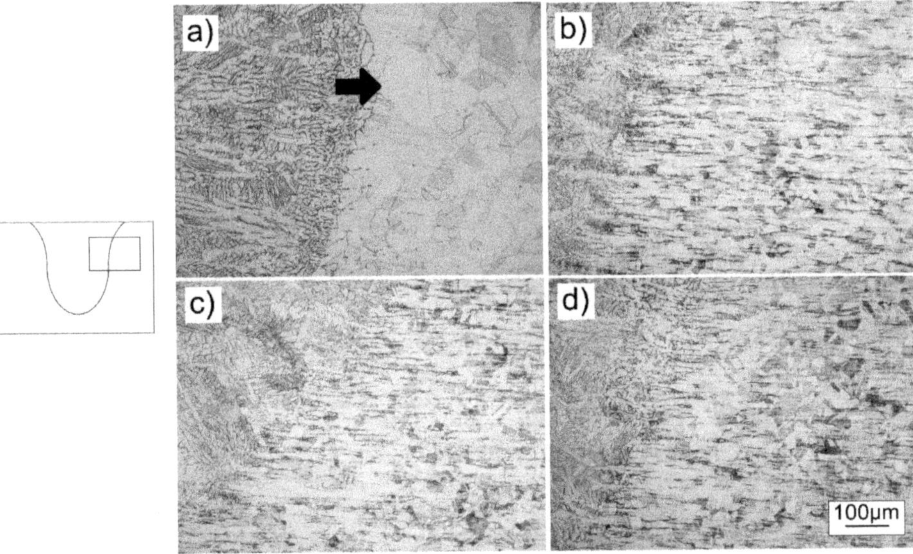

**Figure 4.** Microstructures under the surface of the specimen near the melt line: (**a**) specimen 0; (**b**) specimen 5Ti; (**c**) specimen 1Si4Ti; (**d**) specimen 2Si3Ti.

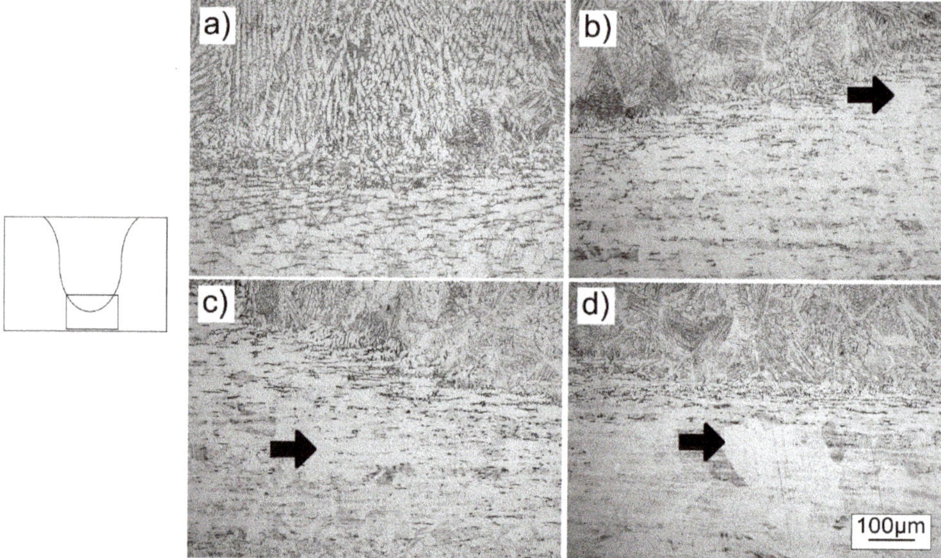

**Figure 5.** Microstructures under the weld metal near the melt line: (**a**) specimen 0; (**b**) specimen 5Si; (**c**) specimen 4Si1Ti; (**d**) specimen 2.5Si2.5Ti.

Microstructures in the base metal near the melt line under the surface of the material are shown in Figure 4. Specimens obtained without the coating (specimen 0) and with the coating (specimens 5Ti, 1Si4Ti, and 2Si3Ti) are shown with significant differences. The specimen obtained without the coating had significantly larger austenite polygonal grains in this area (Figure 4a, indicated by an arrow), which were not present in specimens obtained with the coating (Figure 4b–d).

Differences between microstructures under the weld of specimens 0, 5Si, 4Si1Ti, and 2.5Si2.5Ti are shown in Figure 5. It can be seen that, as in Figure 4, certain differences existed, consisting mainly of coarser grains observed in specimens obtained with the coating (Figure 5b–d, indicated by arrows) as compared to the specimen obtained without the coating (Figure 5a). Note that such a microstructure was not obtained directly under the weld metal, but rather slightly to the side.

*3.4. Vickers Microhardness Results*

The microhardness results are shown in Figures 6 and 7, while average values of the microhardness data measured in the weld metal are shown in Table 1. It can be seen that the major difference between welds obtained without (Figure 6a) and with the nanoparticle-based coating (Figure 6b–h) was in the microhardness values of the weld metals, as shown in Table 1. This was due to the application of a consumable welding wire. Additionally, based on the results shown in Figures 6 and 7 it can be seen that there were drops in microhardness occurring in the base metal near the melt line. However, there were differences between specimens obtained without and with the coating. Namely, in the specimen welded without the coating (0) drops in microhardness occurred in Line 1, right next to the weld line (Figure 6a, marked with black arrows). In specimens welded with coatings (Figure 6b–h) no such drop in microhardness was noted, but a similar drop in microhardness was observed right under the bottom of the weld (Figure 6b–h), which was not observed in the specimen welded without the coating (Figure 6a). The increased microhardnesses obtained in the weld metal were the result of the application of a consumable material in the form of a welding wire.

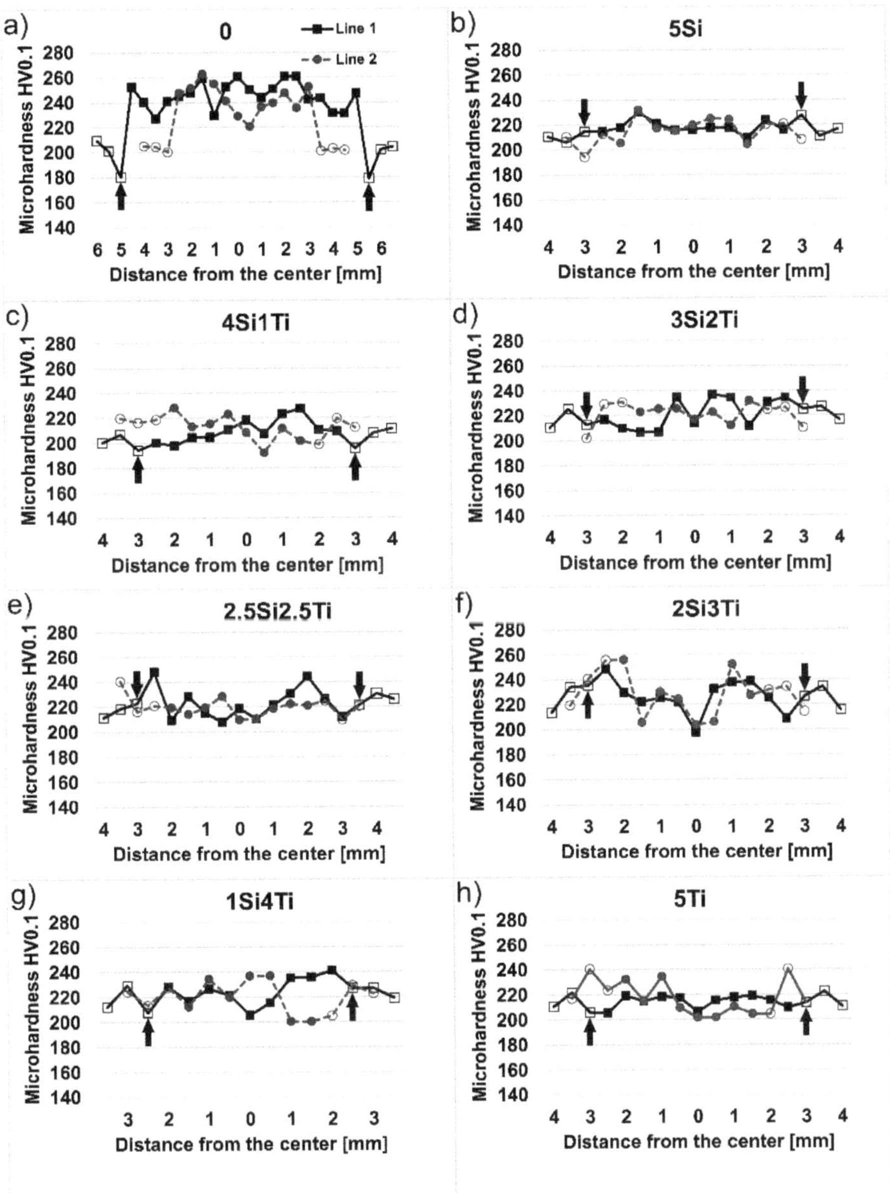

**Figure 6.** Microhardness values measured in Line 1 and Line 2: (**a**) specimen 0; (**b**) specimen 5Si; (**c**) 4Si1Ti; (**d**) 3Si2Ti; (**e**) 2.5Si2.5Ti; (**f**) 2Si3Ti; (**g**) 1Si4Ti; (**h**) 5Ti.

**Table 1.** Average microhardness values for Lines 1, 2, and 3 measured in the weld metal.

| Measurement Area | 0 | 5Si | 4Si1Ti | 3Si2Ti | 2.5Si2.5Ti | 2Si3Ti | 1Si4Ti | 5Ti |
|---|---|---|---|---|---|---|---|---|
| Line 1 | 246 | 218 | 210 | 221 | 223 | 226 | 225 | 214 |
| Line 2 | 243 | 218 | 212 | 222 | 218 | 225 | 220 | 214 |
| Line 3 | 245 | 215 | 214 | 220 | 221 | 227 | 223 | 212 |

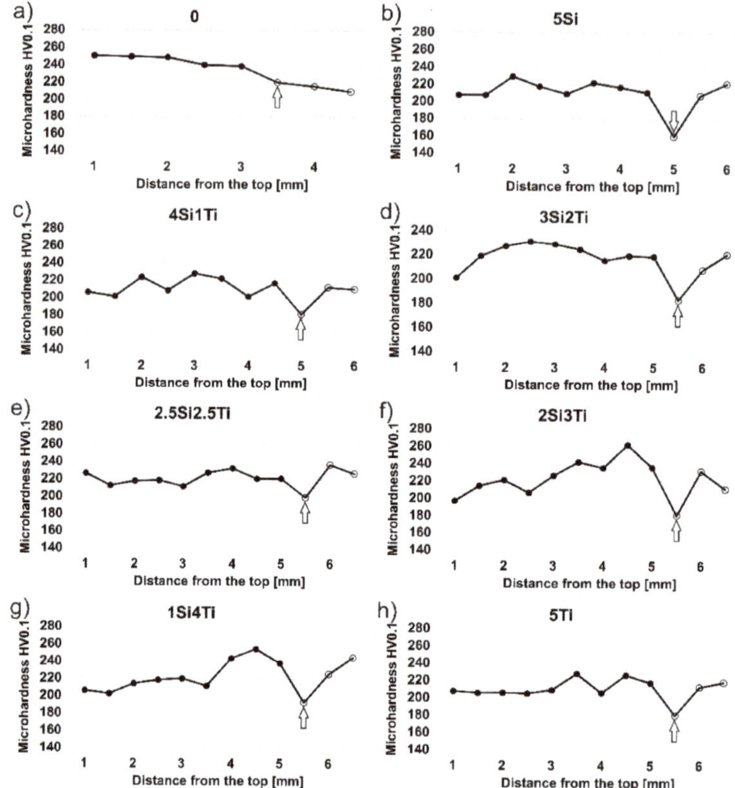

**Figure 7.** Microhardness values measured in Line 3: (**a**) specimen 0; (**b**) specimen 5Si; (**c**) 4Si1Ti; (**d**) 3Si2Ti; (**e**) 2.5Si2.5Ti; (**f**) 2Si3Ti; (**g**) 1Si4Ti; (**h**) 5Ti.

## 4. Discussion

The application of a nanoparticle-based coating applied before the TIG welding had multiple effects. The first and most important was the increase in weld penetration, even though the welding without the coating applied was used with a welding wire and a relatively small V-groove. The increase in penetration was 61% in the case of the most effective mixture containing 2% $SiO_2$ and 3% $TiO_2$ nanoparticles, closely followed by 1% $SiO_2$ and 4% $TiO_2$ nanoparticles. Unary $TiO_2$ coatings proved clearly superior to $SiO_2$-based coatings. These results were not in agreement with the results obtained in studies by Venkatesan et al. [22,31], who found that a $SiO_2$-based coating was more effective compared to a $TiO_2$-based coating at the same welding parameters. The results obtained by Venkatesan et al. [22,31] could be explained by a higher electronegativity of Si (Pauling scale 1.9) versus Ti (Pauling scale 1.54) [20,32,33]. However, one possible explanation in the discrepancy between this study and studies by Venkatesan et al. [22,31] is the size of oxide particles in the coating. Namely, when mixed with the solvent (i.e., acetone), the results suggested that particle agglomeration of $TiO_2$ was more favorable towards a smaller size compared to $SiO_2$. A higher amount of relatively small particles in the flux can be well correlated to a higher effectiveness of a $TiO_2$-based coating compared to a $SiO_2$ one. Microparticles have a lower tendency to agglomerate [34], and therefore electronegativity in A-TIG welding on penetration depth can be of greater importance.

The most convenient particle size distribution was obtained with mixtures of $SiO_2$ and $TiO_2$ nanoparticles, most particularly 2Si3Ti, which closely corresponded to the highest penetration of the specimen obtained with the application of this particular coating. These results are in accordance with the results obtained by Venkatesan et al. [22,31], who also obtained the highest penetration not with single-component coatings, but rather with mixtures—particularly the $SiO_2$ and $TiO_2$ metal-oxides, which were also used in our study.

When nanoparticles are used, as in the case of the present study, it must be noted that the nanoparticle nominal size cannot be considered as influential. Rather the distribution of nanoparticles in accordance to their size is of utmost importance, which accounts for the occurrence of nanoparticle agglomeration, the sizes of which were measured by a Zetasizer in the present study. The results observed with nanoparticle mixtures could be the result of a lower tendency to agglomerate—that is, a higher amount of relatively small particles agglomerates with size up to 1 µm. A similar result regarding a lower tendency to form agglomerates with nanoparticle mixtures was obtained by Efimov et al. [35].

The results of penetration suggest a significant advantage of the application of coatings for TIG welding versus the preparation with an applied consumable wire. The main mechanism of the increased penetration depth was the reversal of the Marangoni convection, proved by the microstructural study. The material flow model with molten metal temperature indicators are shown in Scheme 2. In the specimen welded without the coating (specimen 0), the material flowed from the center of the molten surface towards the melting lines and then towards the bottom and back to the centerline. In this process, the heat was dissipated mainly to the base metal on the sides. This means that the highest heat transfer was towards this area, which caused recrystallization and grain growth, as documented by the microstructure shown in Figure 4a (recrystallization in the horizontal plane). The unchanged microstructure under the weld suggests that the heat transfer towards this area was much less intensive, which is understandable since the melt dissipated its heat to the sides, next to the weld. This was proved by the microhardness measurements with minimum values obtained in this area (Figure 6a). On the other hand, a reversed Marangoni convection caused the material to flow inwards and towards the bottom of the weld, increasing penetration, as well as heat transfer under the weld into the base metal. This means that the area under the weld received the most heat, which resulted in the austenite grain growth—that is, recrystallization in the vertical plane (Figure 5b–d). Similarly to the specimen obtained without the coating, specimens with the applied coating showed a marked drop in microhardness under the weld (Figure 7b–h).

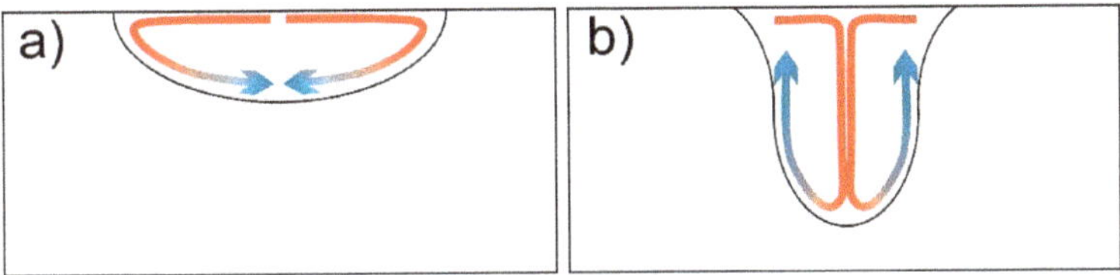


**Scheme 2.** Material flow model: (**a**) without the coating; (**b**) with the coating.

Recrystallization in the horizontal plane that occurred in TIG without the coating was less convenient, weakening the material perpendicular to the weld, unlike the recrystallization in the vertical plane in A-TIG with a truly multifunctional coating. This means that three crucial benefits are obtained with the application of coatings:

1. Improved penetration, along with all accompanying benefits of a lower shielding gas consumption, an increased productivity, a lower environmental impact, and lower costs;
2. A lower negative influence on the material's microstructure and subsequently higher tensile strength, which is proportional to the hardness of the material;
3. A completely eliminated consumable material (i.e., welding wire) containing CRMs such as silicon and a near-CRM such as chromium.

However, special attention must be paid to the root of the weld, keeping in mind that the material flow and its heat transfer was to the bottom part of the weld metal, reducing viscosity and increasing the tendency of the molten metal to flow, causing over-penetration.

## 5. Conclusions

Based on the results presented in this work, the following conclusions can be drawn:

- Coatings containing metallic oxide nanoparticles offer a higher A-TIG welding penetration compared to the specimen without the coating;
- The added consumable material in TIG results in a higher hardness of the weld metal compared to the specimens that were re-melted—that is, welded without the consumable material (A-TIG);
- The penetration of A-TIG specimens depends primarily on the particle size in the coating solution, not the primary particle size, due to the agglomeration of nanoparticles;
- smaller size of agglomerates in coatings with an increased $TiO_2$ content caused a higher coating effectiveness, leading to an increased penetration that reached a value of 61% over that of the TIG weld without the coating;
- The presence of the reversed Marangoni convection was proved by the recrystallization of certain areas of the base metal near the melt line: near the surface in TIG specimen and under the weld in A-TIG specimens. These results were proved by microhardness measurements, with marked drops in microhardness values in recrystallized areas.

An A-TIG coating based on metallic oxide nanoparticles acting as catalysts for welding process gives several benefits to the resulting weld: improves technological properties and improves environmental friendliness (higher penetration, lower shielding gas consumption, improved productivity, etc.), mechanical integrity (weak parts in A-TIG specimens are theoretically under the weld rather than to the sides) and finally, a complete elimination of CRMs and near-CRMs such as Si and Cr, respectively. On the other hand, the catalytic activity of the coating may cause over-penetration as the high-temperature molten metal flows directly downwards, towards the weld root.

**Author Contributions:** S.B. designed the experiment and wrote the paper; M.D. and P.J. performed the experiments; I.Z. provided the resources (devices, materials); D.K. and M.B. interpreted the data, and reviewed and edited the manuscript; M.L.G. provided work administration and supervision.

**Funding:** This research received no external funding.

**Acknowledgments:** This article is based on work from COST Action "Solutions for Critical Raw Materials under Extreme Conditions", supported by COST (European Cooperation in Science and Technology).



## References

1. Gulicovski, J.J.; Bajat, J.; Miskovic-Stankovic, V.; Jokic, B.; Panic, V.; Milonjic, S. Cerium oxide as conversion coating for the corrosion protection of aluminum. *J. Electrochem. Sci. Eng.* **2017**, *3*, 151–156. [CrossRef]
2. Bieber, J. Trivalent chrome conversion coating for zinc and zinc alloys. *Mater. Lett.* **2007**, *105*, 425–435.
3. Ana-Maria, L.; Wolfgang, P.Y.; Sabrina, M.; Nadine, P.; Diane, S.; Tendero, C.; Vahlas, C. Corrosion protection of 304L stainless steel by chemical vapour deposited alumina coatings. *Corros. Sci.* **2014**, *81*, 125–131.
4. Cellarda, A.; Garniera, V.; Fantozzia, G.; Baretb, G.; Fort, P. Wear resistance of chromium oxide nanostructured coatings. *Ceram. Int.* **2009**, *35*, 913–916. [CrossRef]
5. Hwanga, B.; Leea, S.; Ahn, J. Effect of oxides on wear resistance and surface roughness of ferrous coated layers fabricated by atmospheric plasma spraying. *Mat. Sci. Eng. A* **2002**, *335*, 268–280. [CrossRef]
6. Fomin, A.A.; Rodionov, I.V. *Handbook of Nanoceramic and Nanocomposite Coatings and Materials*, 1st ed.; Butterworth-Heinemann: Oxford, UK, 2015; pp. 403–424.
7. Curran, J.A.; Clyne, T.W. The thermal conductivity of plasma electrolytic oxide coatings on aluminium and magnesium. *Surf. Coat. Technol.* **2005**, *199*, 177–183. [CrossRef]

8.   Narottam, P.B.; Dongming, Z. Thermal Properties of Oxides with Magnetoplumbite Structure for Advanced Thermal Barrier Coatings. *Surf. Coat. Technol.* **2008**, *202*, 2698–2703.

9.   Calnan, S. Applications of Oxide Coatings in Photovoltaic Devices. *Coatings* **2014**, *4*, 162–202. [CrossRef]

10.  Sang, Y.J.; Jaesun, S.; Sanghan, L. Photoelectrochemical Device Designs toward Practical Solar Water Splitting: A Review on the Recent Progress of BiVO4 and BiFeO3 Photoanodes. *Appl. Sci.* **2018**, *8*, 1388–1404.

11.  Thakur, P.P.; Chapgaon, A.N. A Review on Effects of GTAW Process Parameters on weld. *IJRASET* **2016**, *4*, 136–140.

12.  German Development Service-GIZ, Gas Tungsten Arc Welding-GTAW. 2000. Available online: https://www.giz.de/expertise/downloads/Fachexpertise/en-metalwork-gas-tungsten-arc-welding.pdf (accessed on 15 March 2019).

13.  Cibi, A.J.; Thilagham, K.T. High Frequency Gas Tungsten Arc Welding Process for Dressing of Weldment. *IJAERS* **2017**, *4*, 229–235.

14.  Robert, W.; Messler, J. *Principles of Welding: Processes, Physics, Chemistry, and Metallurgy*; Wiley-VCH: Weinheim, Germany, 2004; p. 55.

15.  Ramkumar, K.D.; Goutham, P.S.; Radhakrishna, V.S.; Tiwari, A.; Anirudh, S. Studies on the structure–property relationships and corrosion behaviour of the activated flux TIG welding of UNS S32750J. *Manuf. Process.* **2016**, *23*, 231–241. [CrossRef]

16.  Yangchuan, C.; Zhen, L.; Zunyue, H.; Yida, Z. Effect of cerium oxide flux on active flux TIG welding of 800 MPa super steel. *J. Mater. Process. Technol.* **2016**, *230*, 80–87.

17.  Vora, J.J.; Badheka, J.B. Experimental investigation on mechanism and weld morphology ofactivated TIG welded bead-on-plate weldments of reduced activationferritic/martensitic steel using oxide fluxes. *J. Manuf. Process.* **2015**, *20*, 224–233. [CrossRef]

18.  Jurica, M.; Kozuh, Z.; Garasic, I.; Busic, M. Optimization of the A-TIG welding for stainless steels. *IOP Conf. Ser. Mater. Sci. Eng.* **2018**, *329*, 1–9. [CrossRef]

19.  Klobcar, D.; Tusek, J.; Bizjak, M.; Simoncic, S.; Lešer, V. Active flux tungsten inert gas welding of austenitic stainless steel AISI 304. *METABK* **2016**, *55*, 617–620.

20.  Vidyarthy, R.S.; Dwivedi, D.K. Activating flux tungsten inert gas welding for enhanced weld penetration. *J. Manuf. Process.* **2016**, *22*, 211–228. [CrossRef]

21.  Vora, J.J.; Badheka, J.B. Improved Penetration with the Use of Oxide Fluxes in Activated TIG Welding of Low Activation Ferritic/Martensitic Steel. *Trans. Indian Inst. Met.* **2016**, *69*, 1755–1764. [CrossRef]

22.  Venkatesan, G.; George, J.; Sowmysari, M.; Muthupandi, V. Effect of ternary fluxes on depth of penetration in A-TIG welding AISI 409 ferritic stainless steel. *Procedia Mater. Sci.* **2014**, *5*, 2402–2410. [CrossRef]

23.  Takeuchi, Y.; Takagi, R.; Shinoda, T. Effect of bismuth on weld joint pen-etration in austenitic stainless steel. *Weld. Res. Suppl.* **1992**, *71*, 283–290.

24.  Mills, K.C.; Keene, B.J.; Brooks, R.F.; Shirali, A. Marangoni effects in welding. *Philos. Trans. R. Soc. A Math. Phys. Eng. Sci.* **1998**, *356*, 911–925. [CrossRef]

25.  Skvortsov, E.A. Role of electronegative elements in contraction of thearc discharge. *Weld Int.* **1998**, *12*, 471–475. [CrossRef]

26.  Tanaka, M.; Shimizu, T.; Terasaki, T.; Ushio, M.; Koshiishi, F.; Terasaki, H. Effects of activating flux on arc phenomena in gas tungstenarc welding. *Sci. Technol. Weld Join.* **2000**, *5*, 397–402. [CrossRef]

27.  Kuang-Hung, T.; Po-Yu, L. UNS S31603 Stainless Steel Tungsten Inert Gas Welds Made with Microparticle and Nanoparticle Oxides. *Materials.* **2014**, *7*, 4755–4772.

28.  International Standard ISO 9692-1:2013. *Welding and Allied Processes—Types of Joint Preparation—Part 1: Manual Metal-arc Welding, Gas-Shielded Metal-arc Welding, Gas Welding, TIG Welding and Beam Welding of Steels*; ISO: Geneva, Switzerland, 2013.

29.  EUROPEAN TYRE & RUBBER MANUFACTURERS' ASSOCIATION. Available online: http://www.etrma.org/uploads/Modules/Documentsmanager/20170913---2017-list-of-critical-raw-materials-for-the-eu.pdf (accessed on 24 November 2018).

30.  Report on Critical Raw Materials for EU, Report of the Ad-Hoc Working Group on Defining Critical Raw Materials for EU. May 2014. Available online: http://mima.geus.dk/report-on-critical-raw-materials_en.pdf (accessed on 24 November 2018).

31.  Venkatesan, G.; Muthupandi, V.; Justine, J. Activated TIG welding of AISI 304L using mono- and tri-component fluxes. *Int. J. Adv. Manuf. Technol.* **2017**, *93*, 329–336. [CrossRef]

32. Housecroft, C.E.; Sharpe, A.G. *Inorganic Chemistry*, 2nd ed.; Pearson, Prentice Hall, Pearson Education Limited: Edinburgh, UK, 2005; p. 38.
33. Tseng, K.H.; Chen, K.L. Comparisons between $TiO_2$- and $SiO_2$-flux assisted TIG welding processes. *J. Nanosci. Nanotechnol.* **2012**, *12*, 6359–6367. [CrossRef] [PubMed]
34. Trunec, M. Dispersion of nanoparticles in solvents, polymer solutions and melts, principles and possibilities. In Proceedings of the COST Action MP0701 Workshop, Novi Sad, Serbia, 23–24 September 2010.
35. Efimov, A.; Lizunova, A.; Sukharev, V.; Ivanov, V. Synthesis and Characterization of $TiO_2$, $Cu_2O$ and $Al_2O_3$ Aerosol Nanoparticles Produced by the Multi-Spark Discharge Generator. *Korean J. Mater. Res.* **2016**, *26*, 123–129. [CrossRef]

*Article*

# Study of the Deposition Formation Mechanism in the Heat Exchanger System of RHF

**Yuzhu Pan, Xuefeng She *, Jingsong Wang and Yingli Liu**

State Key Laboratory of Advanced Metallurgy, University of Science and Technology Beijing (USTB),
30 Xueyuan Road, Haidian District, Beijing 100083, China; panyuzhuustb@163.com (Y.P.);
wangjingsong@ustb.edu.cn (J.W.); liuyingli112@126.com (Y.L.)
* Correspondence: shexuefeng@ustb.edu.cn; Tel.: +86-138-1156-9346

Received: 22 March 2019; Accepted: 13 April 2019; Published: 15 April 2019

**Abstract:** In rotary hearth furnace (RHF) production, the heat transfer system will produce deposition, which blocks the exhaust channel. The formation of deposition will affect RHF production. In this study, the formation mechanism of deposition was determined through chemical composition analysis, XRD and SEM-EDS: the main cohered phase in the deposition was KCl and the secondary cohered phase was $ZnFe_2O_4$; the $ZnFe_2O_4$ had become solid since it was formed with a porous structure and it cohered other substances; the $ZnFe_2O_4$ exhibited stronger cohering strength than KCl, due to a different cohering mechanism. In contrast, the KCl played a significant role in the deposition on the heat exchanger wall. A new process was proposed to avoid the deposition formation. This process could eliminate the deposition in the heat transfer system of RHF and improve the utilization of metallurgical waste.

**Keywords:** RHF; deposition; KCl; $ZnFe_2O_4$; cohering mechanism

## 1. Introduction

Large amounts of waste are generated in ironmaking and steelmaking processes, such as sinter dusts, blast furnace dusts and converter dusts. These wastes are valuable because they contain Fe and C. Adversely, these dusts are not directly used in blast furnace production because they contain high levels of Pb, Zn, K and Na, which are harmful to blast furnace production. The direct reduction of the ironmaking process of rotary hearth furnace (RHF) is an effective method for the treatment of metallurgical wastes and it has received significant attention [1–4]. Five RHFs for the treatment of metallurgical wastes have existed in Kimtsu, Hikari and Hirohata in Japan since 2000 [5,6]. In China, the Laiwu Steel, the Maanshan Iron and Steel, the Shangan, the Rizhao Iron and Steel and other enterprises have built RHFs for dealing with metallurgical wastes. Usually, the wastes are turned into carbon-bearing pellets, from which C is effectively utilized for the reduction of metal elements. Pb and Zn through the reduction and re-oxidation into the gas phase form the secondary dust. KCl will volatilize into the gas phase and Fe will form direct reduced iron (DRI). Zn, Pb and KCl are obtained through the secondary dust treatment. The DRI can be used for Blast furnace process as confirmed by She [7]. Through the RHF process, the separation of harmful elements and Fe is achieved, in order for the metallurgical wastes to be effectively utilized.

In order to improve the RHF energy efficiency, the gas waste heat was recovered. In the waste heat recovery process, Zn, Pb, K and Fe produce deposition, which blocks the exhaust channel. The production of RHF is affected, because the occurrence of deposition reduces the heat transfer efficiency and increases the furnace pressure. In the past, regarding the RHF process study, researchers were often concerned about the metal element reduction and energy efficiency [8–12], while the effect of deposition on the heat transfer system for RHF was ignored. In this study, the deposition was analyzed through chemical analysis, XRD, as well as SEM-EDS analysis and experiments. Consequently, the formation mechanism of deposition was obtained.

## 2. Experimental Procedure

The deposition sample was obtained from the RHF waste heat recovery system, as presented in Figure 1. In this RHF, sintering dust, pellet dust and blast furnace dust were utilized as raw materials to produce metalized pellets. The RHF production cycle was 15 min. The RHF production is presented in Figure 2. The waste heat recovery system consisted of a high temperature heat exchanger and a low temperature heat exchanger. The inlet temperature of the high temperature heat exchanger was approximately 1200 °C, the outlet temperature was approximately 700 °C, while the flue gas flow rate was 5–6 × 10$^3$ m$^3$/h. The deposition mainly occurred at the wall end of the high temperature heat exchanger.

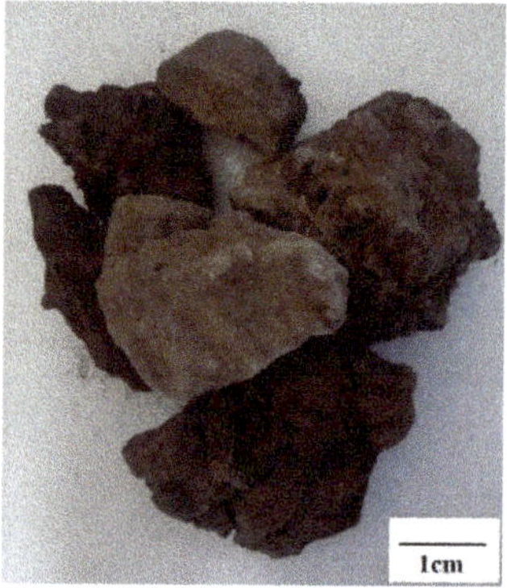


**Figure 1.** Deposition samples of RHF waste heat recovery system.

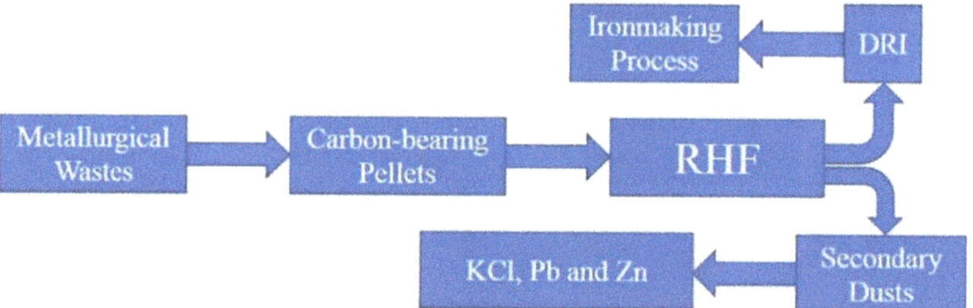


**Figure 2.** Waste treatment of RHF process.

In order to determine the deposition formation mechanism, the deposition samples were analyzed through composition analysis, XRD and SEM-EDS, while the determination of cohered phase experiments were carried out according to the composition and XRD analyses. The results of composition analysis and XRD are presented in Table 1 and Figure 3, respectively. The conducted experiments are presented in Table 2. The main elements in the deposition are Zn, Pb, K and Fe. During the design experiment, Zn, Pb, Fe and K are expressed in the form of ZnO, PbO, Fe$_2$O$_3$ and KCl according to the actual elements' ratio of deposition. However, some of the Zn elements exist in the form of ZnCl$_2$ and other minor components such as CaO, SiO$_2$ and C are present, which may slightly affect the composition reported in Table 2. The raw materials were pure chemical reagents. The weight of each raw material was 20 g and the raw materials were thoroughly mixed. Mixed raw materials were placed in a corundum crucible and heated in a muffle furnace for 15 min. The experimental temperature was 700 °C. The purpose of experiments 1–4 was to determine the cohered phases, whereas their composition was designed according to the compositions of Table 1. The samples 1–4 were tested with XRD. According to Figure 3, the deposition structure consisted of five simple compounds and a complex compound. Therefore, it was necessary to study the effect of $ZnFe_2O_4$ for the deposition formation. The experiment 5, in order to study the effect of $ZnFe_2O_4$ on the deposition formation, was carried out twice: one of the samples was subjected to water quenching, while the other sample was cooled at room temperature. The two samples were analyzed through XRD and SEM-EDS. A high gas flow rate existed in heat exchanger system of RHF. The deposition product must have the ability to resist producing mechanical damage to the wall stability of the heat exchanger. Therefore, a drop strength test was designed to detect the bond strength of various substances. Subsequently to the sample obtained in experiment 8–13 room temperature cooling, the sample was allowed to fall into a sufficiently large corundum crucible at a height of 20 cm. Following the sample drop, the powder was collected and the bond strength of the sample was $N$ times/20 cm, subsequently to falling $N$ times when the powder exceeded 50% of the total mass.

**Table 1.** Chemical composition of deposition (mass %).

| Chemical Compositions % | CaO | SiO$_2$ | Zn | Pb | K | C | TFe |
|---|---|---|---|---|---|---|---|
| Deposition | 1.62 | 1.08 | 36.5 | 8.4 | 11.57 | 0.66 | 12.84 |

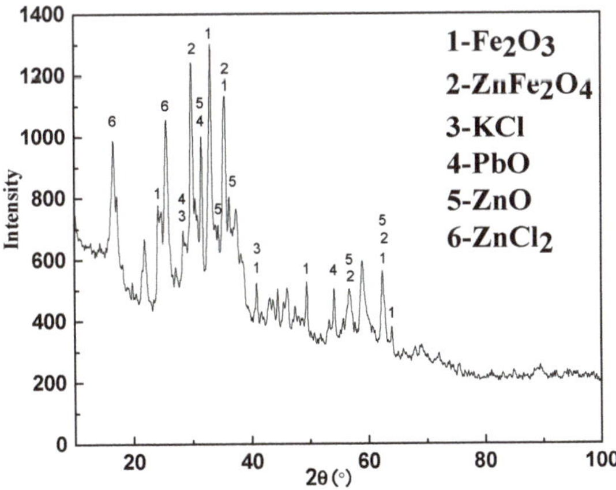


**Figure 3.** XRD pattern of deposition products.

**Table 2.** Chemical composition of experimental materials (mass %).

| Experiments | ZnO | PbO | Fe$_2$O$_3$ | KCl |
|---|---|---|---|---|
| 1 | 84.38 | 15.62 | - | - |
| 2 | 62.89 | 11.63 | 25.48 | - |
| 3 | 63.94 | 11.83 | - | 24.22 |
| 4 | 50.81 | 9.4 | 20.53 | 19.26 |
| 5 | 50.31 | - | 49.46 | - |
| 6 | 80.78 | 14.22 | - | 5 |
| 7 | 76.43 | 13.57 | - | 10 |
| 8 | 72.29 | 12.71 | - | 15 |
| 9 | 67.86 | 12.14 | - | 20 |
| 10 | 80.78 | 14.22 | 5 | - |
| 11 | 76.43 | 13.57 | 10 | - |
| 12 | 72.29 | 12.71 | 15 | - |
| 13 | 67.86 | 12.14 | 20 | - |

## 3. Results and Discussion

### 3.1. Analysis Results of Deposition

The chemical composition of the deposition and the XRD results are given in Table 1 and Figure 3. According to Table 1 and Figure 3, it could be observed that the deposition product was mainly composed of ZnO, PbO, KCl, $Fe_2O_3$, $ZnCl_2$ and $ZnFe_2O_4$. The ZnO and PbO were derived from the reduction and re-oxidation of Zn and Pb in the carbon-bearing pellets. The KCl was volatilized directly from the carbon-bearing pellets in the high temperature furnace of the RHF. The $ZnCl_2$ was not present in the carbon-bearing pellets, while its formation was more complicated. The Cl in $ZnCl_2$ was derived from KCl. The key to the formation of $ZnCl_2$ was the decomposition of KCl. Certain studies demonstrated that the corrosion of pure iron in oxidizing atmosphere with KCl vapor could result in the following reaction [13–15]:

$$2Fe_2O_3 + 2KCl_{(g)} + \frac{1}{2}O_2 = K_2Fe_4O_7 + Cl_{2(g)} \tag{1}$$

A high amount of $K_2Fe_4O_7$ was detected on the etched iron surface, indicating that the Reaction (1) could separate the K and Cl. Unfortunately, the thermodynamic data of Reaction (1) were not found. The vapor of Zn and $Cl_2$ can react as:

$$Zn_{(g)} + Cl_{2(g)} = ZnCl_{2(g)} \tag{2}$$

The reason for the $ZnCl_2$ formation in the deposition product could be explained by Reactions (1) and (2). The $Fe_2O_3$ was derived from the carbon-bearing pellets. The carbon-bearing pellets would produce cracks or even break to produce high-sized particles of iron-containing dust in the heating and reduction process. Blast furnace dust, which contained a certain amount of $ZnFe_2O_4$, was one of the raw materials for RHF. Yet, the $ZnFe_2O_4$ would be destroyed during the reduction of carbon pellets as presented through Reactions (3)–(5):

$$ZnFe_2O_4 + C = ZnO + 2FeO + CO_{(g)} \tag{3}$$

$$ZnFe_2O_4 + CO_{(g)} = ZnO + 2FeO + CO_{2(g)} \tag{4}$$

$$ZnFe_2O_4 + 2Fe = 4FeO + Zn_{(g)} \tag{5}$$

It could be determined that the $ZnFe_2O_4$ was produced by the contact between $Fe_2O_3$ and ZnO during deposition. The formation mechanism of $ZnFe_2O_4$ and its role in the deposition was described in Sections 3.2 and 3.3.

The typical morphology of the deposition is presented in Figure 4. Apparently flat, porous and transitional areas on the deposition surface existed, with certain bright white areas being distributed on the deposition product surface. In Figure 4, the left side was near the end of the heat exchanger wall and the right side was the end near the center of the heat exchanger.

The EDS maps of the transitional and flat areas are presented in Figure 5. In Figure 5, the sample was taken from the heat exchanger wall surface and the upper part of the sample was attached to the heat exchanger wall. It could be observed that the distribution areas of Cl (Figure 5b) and K (Figure 5c) overlapped, combining the chemical composition of Table 1 and the results of Figure 3. The XRD result demonstrated that the cubic particles of the transitional region were KCl, while the other locations of the transitional areas were filled with ZnO. The flat areas were mainly composed of ZnO particles, while the bright white areas were mainly composed of PbO (Figure 5f).

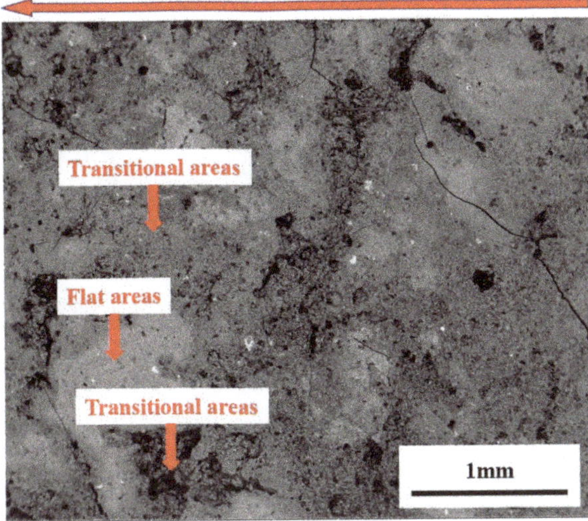



**Figure 4.** Typical morphology of deposition surface.

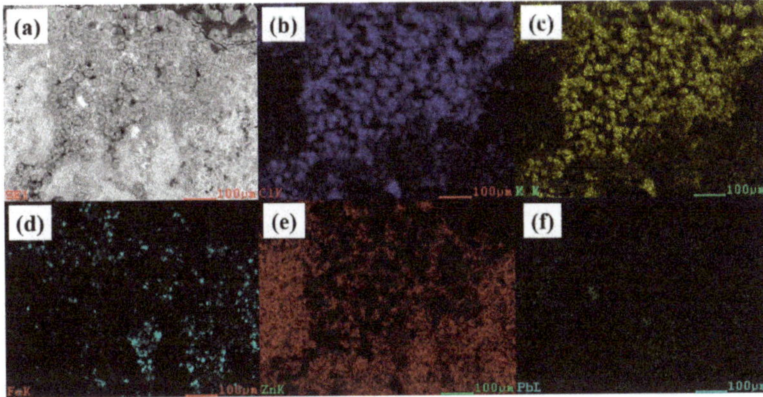



**Figure 5.** EDS maps of flat and transitional areas. (**a**) EDS map; (**b**) Cl; (**c**) K; (**d**) Fe; (**e**) Zn; (**f**) Pb.

Figure 6 presents the EDS maps results for the porous areas. It was also found that the distribution of Cl (Figure 6b) and K (Figure 6c) overlapped, but no cubic KCl particles appeared. No enrichment of PbO was observed in the porous region. It was regrettable that no $ZnFe_2O_4$ phase was found in any area. From the EDS maps results of the various regions, certain areas existed, in which, the Fe and Zn overlapped, while the $ZnFe_2O_4$ phase was most likely to appear at these positions. The reason for the $ZnFe_2O_4$ phase absence was that the $ZnFe_2O_4$ phase was covered with KCl and ZnO.

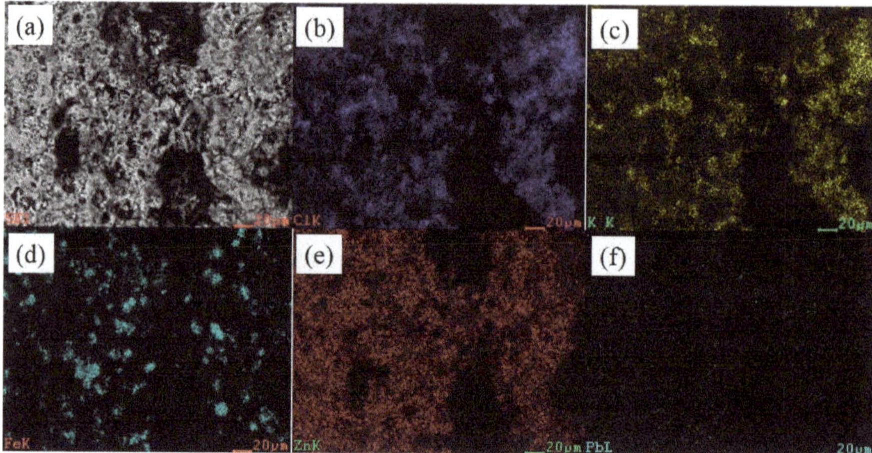

**Figure 6.** EDS maps results for porous areas. (**a**) EDS map; (**b**) Cl; (**c**) K; (**d**) Fe; (**e**) Zn; (**f**) Pb.

Certain filled holes were found on the deposition product surface, as presented in Figure 7a. Following the hole interior enlargement, it was observed that the walls of the hole were covered with molten KCl and cubic KCl particles began to appear (Figure 7b). This was similar to the transitional areas, but these particles were not completely crystallized, compared to the particles of KCl in the flat areas (Figure 7c).

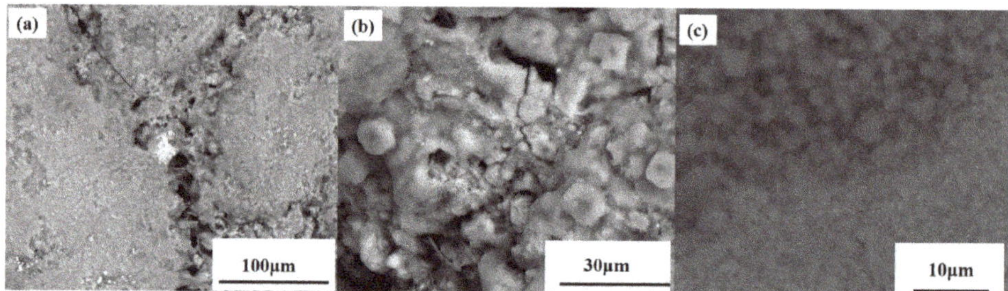

**Figure 7.** Filled holes (**a**), KCl morphology in hole (**b**) and morphology of KCl in flat areas (**c**).

### 3.2. Experimental Results and Discussion

The XRD results of samples of 1–4 groups of experiments in Table 2 are shown in Figure 8. The phases in sample 1 were PbO and ZnO. In addition to the PbO and ZnO phases, $ZnFe_2O_4$ was present in sample 2 due to the addition of $Fe_2O_3$ component to sample 2. Three phases of PbO, ZnO and KCl were present in sample 3, while in sample 4, four phases of PbO, ZnO, KCl and $ZnFe_2O_4$ were present. In the 1–4 groups of experiments, the sample 1 was still powdered following cooling, while the samples in groups 2–4 were bulk after the samples were cooled. Through the XRD results, it could be observed that the mixture of ZnO and PbO did not produce any new compound subsequently to heating, signifying that no cohered phase existed. Therefore, the sample of experiment 1 was still powdery following cooling. The $ZnFe_2O_4$ phase appeared in the sample of experiment 2, while it was possible that the $ZnFe_2O_4$ would bind the other powders together. In experiment 3, although no new compound appeared, the melting point of KCl was low, while the particles of the other substances were adhered to each other during the solidification of KCl. Also, sample 4 presented a result of the combined effect of KCl and $ZnFe_2O_4$. From the 1–4 group of experiments, KCl and $ZnFe_2O_4$ produced the effect of bonding.

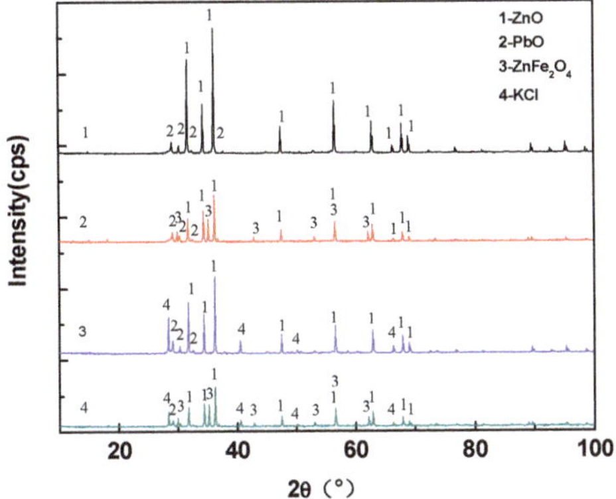



**Figure 8.** Experimental 1–4 sample XRD results.

Figure 9 presents the XRD and SEM results of the two samples in experiment 5, where the water quenched sample was denoted as W and the sample cooled in the air was denoted as A. From the XRD results of the two samples, the ZnFe$_2$O$_4$ could be completely crystallized under the conditions of rapid cooling in water or slow cooling in the air. This result indicated that the ZnFe$_2$O$_4$ was solid when it formed and that the melting point of ZnFe$_2$O$_4$ exceeded 1200 °C. Therefore, the mechanism of ZnFe$_2$O$_4$ formation reaction was obtained, as presented in Equation (6). Through the FactSage 7.0 thermodynamics software, the $\Delta G_{ZnFe2O4}$ was calculated, as presented in Equation (7):

$$ZnO_{(S)} + Fe_2O_{3(S)} = ZnFe_2O_{4(S)} \tag{6}$$

$$\Delta G_{(ZnFe_2O_4)} = -22723.72902 + 2.90446T \ (873\,\text{K} - 1523\,\text{K}) \tag{7}$$

From Equation (7), it could be observed along with Reaction (6) for the endothermic reaction, that the lower temperature was more conducive to the ZnFe$_2$O$_4$ formation. It can be seen from Figure 3 that there was Fe$_2$O$_3$ phase in the actual deposition, but there was no Fe$_2$O$_3$ phase in the sample 5. This may be due to the complete conversion of Fe$_2$O$_3$ to ZnFe$_2$O$_4$ during the formation of actual deposition: (1) the contact probability of Fe$_2$O$_3$ and ZnO particles was lower in the gas phase prior to deposition; (2) Particle of Fe$_2$O$_3$ and ZnO can be stably contacted after being deposited on the heat exchanger wall, but this will result in decrease of particle temperature. Although $\Delta G_{ZnFe2O4}$ was reduced with decrease of temperature and which was favorable for the formation of ZnFe$_2$O$_4$. However, with the formation of ZnFe$_2$O$_4$, the kinetic conditions of the reaction may be destroyed because ZnFe$_2$O$_4$ will block the contact between Fe$_2$O$_3$ and ZnO, and resulting in the Fe$_2$O$_3$ not being fully converted into ZnFe$_2$O$_4$; (3) the Fe$_2$O$_3$ on the deposition product originated from the high-sized particles of dust, resulting in the Fe$_2$O$_3$ and ZnO particles not to be completely in contact. From the SEM-EDS results, many holes existed on the sample surface, but the holes (black) of sample W were dense but low in size, while the holes of sample A appeared sparse but high in size. This might be due to the fact that the W sample rapidly cooled and prevented the contact of different ZnFe$_2$O$_4$ particles (gray phase in the Figure 9) from increasing in size, while the ZnFe$_2$O$_4$ retained the initial state at the time of generation. The sample cooling rate was slow for different ZnFe$_2$O$_4$ crystal contacts and growths to provide the temperature conditions, while the hole size increased due to the ZnFe$_2$O$_4$ crystal growth. The formation of ZnFe$_2$O$_4$ and the crystallization of the other powder particles were fixed, for the entire sample to appear massive. This was also the reason for the sample 2 massive appearance.

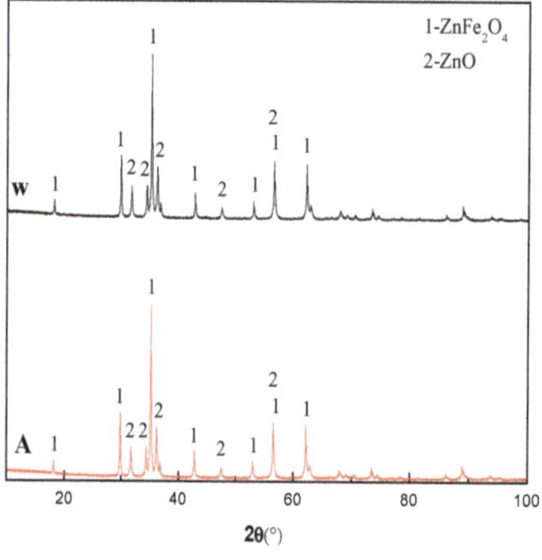 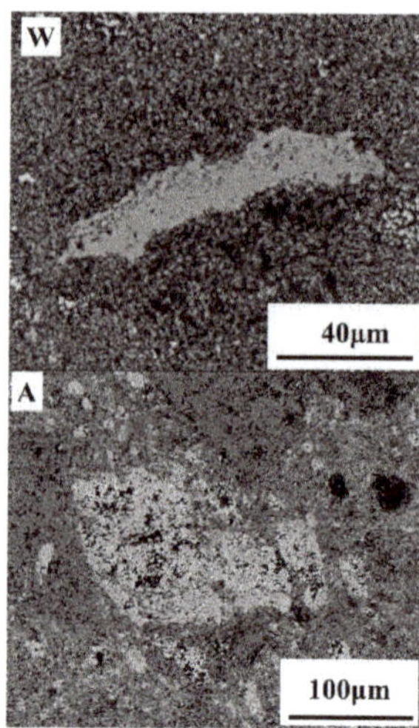



**Figure 9.** Experiment 5 XRD and SEM results of water quenched sample (W) and air cooled sample (A).

The KCl and ZnFe$_2$O$_4$ bond strength test results are presented in Figure 10. The bond strength of the sample increased as the KCl and Fe$_2$O$_3$ contents increased. Among them, since the content of KCl was only 5% in sample 6, the sample itself had unbonded powder. Consequently, the bond strength performance was quite low. From the latter, it could be observed that the cohered phase was the ZnFe$_2$O$_4$, when the Fe$_2$O$_3$ was added to the ZnO and PbO powder matrix. From the experimental results, the bonding effect of ZnFe$_2$O$_4$ was significantly improved compared to KCl. When the KCl content was 15%, the bond strength was 17 times/20 cm, while the ZnFe$_2$O$_4$ was used as the cohered phase. When the Fe$_2$O$_3$ content reached 10%, the bond strength reached 17 times/20 cm. Figure 11 presents the bonding methods of KCl and ZnFe$_2$O$_4$, respectively. This was due to different ZnFe$_2$O$_4$ particle growths to form a macroscopic porous structure, while the other particles were embedded in the hole. Also, in the formation of ZnFe$_2$O$_4$, a part of the bonded phase was consumed as the ZnO changed into a binder phase. In other words, each micro-ZnFe$_2$O$_4$ particle adhered to the solid particles in a lower amount compared to KCl bonded particles. Consequently, the performance of ZnFe$_2$O$_4$ bonding was stronger than KCl. By contrast, in the actual deposition formation, since the formation of ZnFe$_2$O$_4$ was a solid-solid reaction, the material for the deposition immobilization on the wall of the heat exchanger was KCl. The KCl content in the deposition was approximately 22%, whereas when the KCl content reached 20%, the sample bond strength was 30 times/20 cm, indicating that the KCl existence in the deposition was sufficient to fix the deposition on the wall of the heat exchanger. The KCl bonded other phases through the liquid to solid transition, while the ZnFe$_2$O$_4$ produced a porous structure to fix the other solid particles. The ZnFe$_2$O$_4$ was a solid phase that could not self-immobilize along with other substance particles on the heat exchanger walls. Therefore, the KCl was considered as the main cohered phase and the ZnFe$_2$O$_4$ was the secondary cohered phase.

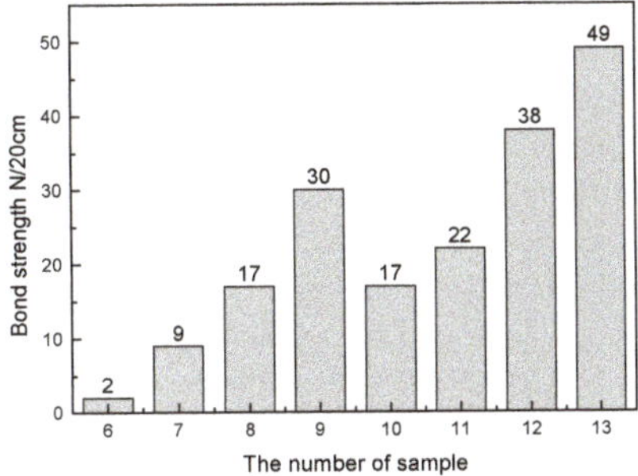

**Figure 10.** Effects of KCl and $Fe_2O_3$ content on sample bond strength.

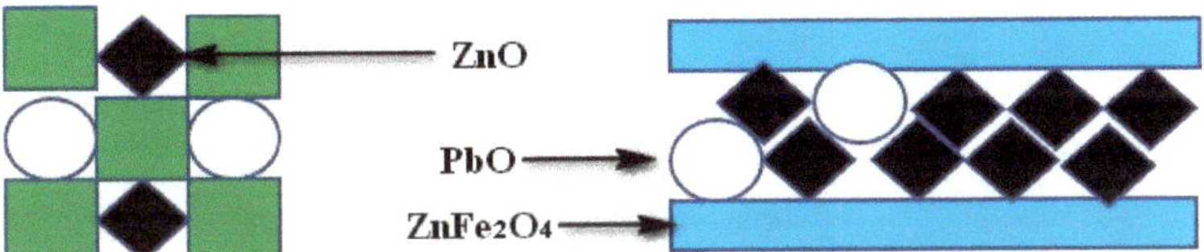


**Figure 11.** Bonding methods of KCl and $ZnFe_2O_4$.

### 3.3. Discussion on Formation Mechanism and Prevention of Deposition

Through Chapters 3.1 and 3.2, two types of cohered phases were identified, namely the KCl and $ZnFe_2O_4$. The bonded phases were ZnO, PbO and the iron-containing particles. Therefore, it could be inferred that the formation mechanism of the deposition product was: (1) the KCl solidified on the heat exchanger wall, while the iron-containing particles, as well as the ZnO and PbO particles would be bonded with KCl during the KCl solidification. Since the iron-containing particle sizes exceeded the ZnO and PbO sizes, the iron-containing particles would remain on KCl earlier than the ZnO and PbO deposition. (2) the $ZnFe_2O_4$ was formed following the contact with ZnO and $Fe_2O_3$, while different $ZnFe_2O_4$ phases grew to form the porous structure. (4) the KCl in the porous structure continued to solidify and bond other dust particles, for the hole to be filled; (5) When the hole was filled, the formations of gray flat and transition areas occurred, not being completely filled, while retaining the porous area morphology. The process of deposition formation is presented in Figure 12.

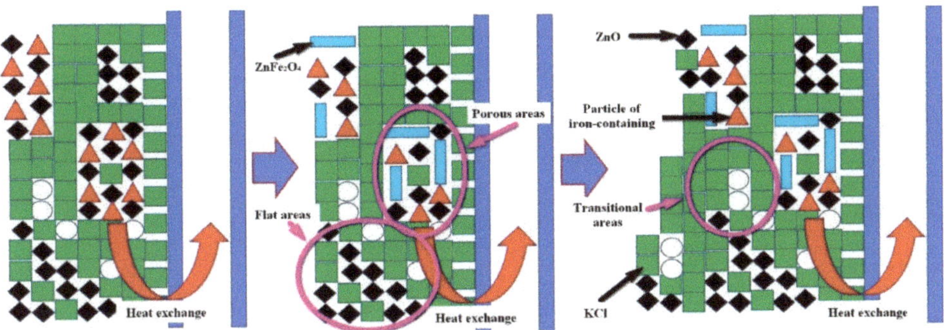


**Figure 12.** Schematic diagram of deposition formation.

In summary, to prevent the formation of deposition product, the appearance of the cohered phase must be avoided. In this study, the cohered phases of the deposition were KCl and $ZnFe_2O_4$. In the production of the rotary hearth furnace, the carbonaceous pellets should be prevented from being dispersed into the flue gas, to cause the iron-containing material and the ZnO to form zinc ferrite. When the rotary hearth furnace was used to treat the metallurgical waste, one of its main functions was to remove and recover both Zn and Pb. Consequently, the presence of ZnO in the flue gas could not be avoided. Therefore, in order to avoid the emergence in the heat transfer system of RHF, it was necessary to avoid the KCl occurrence in the RHF. Guo Zhancheng et al. [16–19] proposed a method of KCl extraction with sintered dust through the sintered dust immersion in water. The KCl dissolved in water during the immersion to form a solution of an insoluble material for the production of iron concentrate, along with a leaching solution for the extraction of KCl. Through the results utilization of Guo Zhancheng et al., a production process to avoid the occurrence of sedimentation was proposed. As presented in Figure 13, the raw material containing KCl was subjected to water immersion treatment prior to pelletizing. Also, the KCl content was maximized within water, producing a leaching solution for the extraction of KCl. Through this process, not only the KCl could be avoided during the production of RHF hazards, but also the maximum use of the waste of the metallurgical process could be achieved.

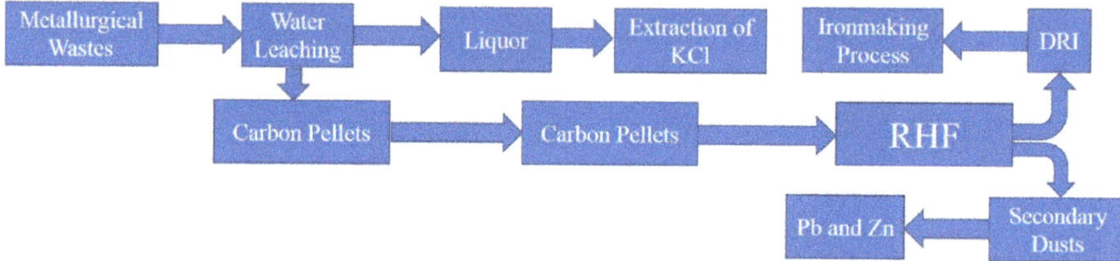


**Figure 13.** A process of RHF without deposition generation.

## 4. Conclusions

Through the detection and experimental study of the deposition in the heat exchanger system, the following conclusions were drawn:

(1)  The main cohered phase of the deposition was KCl. The other solid particles were fixed by KCl through the solid-liquid transition of KCl.

(2)  The secondary cohered phase of the deposition was $ZnFe_2O_4$, forming a porous structure, while the other solid particles were fixed in the porous structure.

(3)  The bond strength of $ZnFe_2O_4$ was better compared to KCl. Under the combined action of KCl and $ZnFe_2O_4$, the deposition could stably occur on the heat exchanger wall.

(4)  A new RHF process was proposed to avoid the generation of sediments and to maximize the use of waste from the metallurgical process.

**Author Contributions:** Writing—original draft, Y.P.; Study and design, J.W.; Date interpretation, X.S.; Literature research and date collection; Y.L.

**Funding:** This research was supported by Natural Science Foundation of China (NO. 51874029), The project of State key laboratory of advanced metallurgy (NO. 41618018) and China Postdoctoral Science Foundation (NO. 2018M641195).

**Conflicts of Interest:** The authors declare no conflict of interest.

## References


1.  She, X.F.; Wang, J.S.; Han, Y.H.; Zhang, X.X.; Xue, Q.G. Comprehensive mathematical model of direct reduction for rotary hearth furnaces. *J. Univ. Sci. Tech. Beijing* **2013**, *35*, 1580–1587.

2.  Peng, C.; Zhang, F.L.; Li, H.F.; Guo, Z.H. Removal behavior of Zn, Pb, K and Na from cold bonded briquettes of metallurgical dust in simulated RHF. *ISIJ Int.* **2009**, *49*, 1874–1881. [CrossRef]

3.  Lee, Y.S.; Ri, D.W.; Yi, S.H.; Sohn, I. Relationship between the reduction degree and strength of DRI pellets produced from iron and carbon bearing wastes using an RHF simulator. *ISIJ Int.* **2012**, *52*, 1454–1462. [CrossRef]

4. Han, H.L.; Duan, D.P.; Yuan, P. Binders and bonding mechanism for RHF briquette made from blast furnace dust. *ISIJ Int.* **2014**, *54*, 1781–1789. [CrossRef]

5. Kawaguchi, T.; Kashiwaya, Y. Preface to the special issue on "recycling of wastes and environmental problems". *ISIJ Int.* **2000**, *40*, 211. [CrossRef]

6. Rutherford, S.D.; Kopfle, J.T. Mesabi Nugget: The world's first commercial ITmk3® plant. *Iron Steel Tech.* **2010**, *7*, 38–43.

7. She, X.F.; Wang, J.S.; Liu, J.Z.; Zhang, X.X.; Xue, Q.G. Increasing the mixing rate of metalized pellets in blast furnace based on the high-temperature interactivity of iron bearing materials. *ISIJ Int.* **2014**, *54*, 2728–2736. [CrossRef]

8. Fu, J.X.; Zhang, C.; Hwang, W.S.; Liau, Y.T.; Lin, Y.T. Exploration of biomass char for $CO_2$ reduction in RHF process for steel production. *Int. J. Greenhouse Gas Control* **2012**, *8*, 143–149. [CrossRef]

9. Wang, G.; Wang, J.S.; Ding, Y.G.; Ma, S.; Xue, Q.G. New separation method of boron and iron from ludwigite based on carbon bearing pellet reduction and melting technology. *ISIJ Int.* **2012**, *52*, 45–51. [CrossRef]

10. Wang, C.L.; Li, K.Q.; Yang, H.F.; Li, C.H. Probing study on separating Pb, Zn, and Fe from lead slag by coal-based direct reduction. *ISIJ Int.* **2017**, *57*, 996–1003. [CrossRef]

11. Fan, D.C.; Ni, W.; Yan, A.Y.; Wang, J.Y.; Cui, W.H. Orthogonal experiments on direct reduction of carbon-bearing pellets of bayer red mud. *J. Iron Steel Res. Int.* **2015**, *22*, 686–693. [CrossRef]

12. Chung, S.H.; Kim, K.H.; Sohn, I. DRI from recycled iron bearing wastes for lower carbon in the blast furnace. *ISIJ Int.* **2015**, *55*, 1157–1164. [CrossRef]

13. He, Y.W.; Li, Y.C.; Zhang, H.L.; Li, M.; Xu, Z.J.; Zhong, F.; Yang, F. High temperature corrosion behavior of T91 in AlKAli metal chloride medium. *Corros. Prot.* **2015**, *36*, 1021–1026. (In Chinese)

14. Ma, H.T.; Guo, G.F.; Zhao, J.; Wang, L. High temperature corrosion behavior of pure iron in atmosphere $O_2$ containing small amount of KCl vapor. *Corros. Sci. Prot. Tech.* **2005**, *17*, 20–23.

15. Li, Y.S.; Niu, Y.; Liu, G.; Wu, W.T. Corrosion of pure Fe and 310SS beneath $ZnCl_2$-KCl salt film at 450 °C. *Acta Metall. Sin.* **2002**, *36*, 1183–1186.

16. Zhang, M.; Fu, Z.G.; Wu, B.; Lv, N.; Zeng, L.L.; Yang, Y.Q. Recovery of potassium chloride from sintering EAF dust. *Chin. J. Process Eng.* **2014**, *14*, 979–983.

17. Zhan, G.; Guo, Z.C. Water leaching kinetics and recovery of potassium salt from sintering dust. *Trans. Nonferr. Met. Soc. China* **2013**, *23*, 3770–3779. [CrossRef]

18. Tang, H.H.; Sun, W.; Han, H.S. A novel method for comprehensive utilization of sintering dust. *Trans. Nonferr. Met. Soc. China* **2015**, *25*, 4192–4200. [CrossRef]

19. Pei, B.; Zhan, G.; Chen, P.Z.; Guo, Z.C.; Gao, J.T. Preparation of potassium chloride and spherical calcium carbonate particles from leaching solution of electrostatic precipitator dust of iron ore sintering. *Chin. J. Process Eng.* **2015**, *15*, 137–146.

*Article*

# Properties of TiC and TiN Reinforced Alumina–Zirconia Composites Sintered with Spark Plasma Technique

**Magdalena Szutkowska *, Sławomir Cygan, Marcin Podsiadło, Jolanta Laszkiewicz-Łukasik, Jolanta Cyboroń and Andrzej Kalinka**

ŁUKASIEWICZ Research Network-IAMT, 37a Wroclawska St., 30-011 Cracow, Poland;
slawomir.cygan@ios.krakow.pl (S.C.); marcin.podsiadlo@ios.krakow.pl (M.P.);
jolanta.laszkiewicz@ios.krakow.pl (J.L.-Ł.); jolanta.cyboron@ios.krakow.pl (J.C.);
andrzej.kalinka@ios.krakow.pl (A.K.)
* Correspondence: szutkows@ios.krakow.pl

Received: 4 October 2019; Accepted: 6 November 2019; Published: 13 November 2019

**Abstract:** In this paper, $Al_2O_3$–$ZrO_2$ composites with an addition of 20 wt% TiN and 10 wt% TiC were modified. The addition of zirconia in a range from 2 to 5 wt% of the monoclinic phase and 10 wt% of $Y_2O_3$ stabilised $ZrO_2$ affected the mechanical properties of the composites. A new type of sintering technique—the spark plasma sintering (SPS) method—within a temperature range from 1575 °C to 1675 °C, was used. Vickers hardness, apparent density, wear resistance and indentation fracture toughness $K_{IC(HV)}$ were evaluated at room temperature. An increase of the sintering temperature resulted in an improvement of Vickers hardness and an increase of the fracture toughness of the tested composites. The tribological properties of the samples were tested using the ball-on-disc method. The friction coefficient was in a range from 0.31 to 0.55, depending on the sintering temperature. An enhancement of the specific wear rate was dependent on the sintering temperature. The mechanical properties of the samples sintered by pressureless sintering (PS) were compared. X-ray diffraction patterns were presented in order to determine the phase composition. SEM microstructure of the tested composites sintered at different temperatures was observed.

**Keywords:** alumina–zirconia composites; TiC; TiN; spark plasma sintering; wear resistance; indentation fracture toughness; X-ray diffraction

---

## 1. Introduction

The technical performance capabilities of traditional tool materials are no longer sufficient to solve many machining problems, so these materials are successively replaced by new ones, including by sintered ceramic materials made with the use of new technologies. Due to the disappearing resources of heavy and high-melting metals and the associated high costs, the growing interest in ceramic matrix composites is also due to economic reasons. It is estimated that ceramics account for 8% of tool materials that are currently being used [1]. Among ceramics, alumina is the most important, widely used and cost-effective oxide ceramic material with perfect properties, such as high thermal resistance, good chemical stability, low density, high hardness and wear resistance, but has relatively low reliability. This is shown by the low fracture toughness $K_{IC}$ and the large dispersion of mechanical properties, reflected by the low values of the Weibull module $m$ [2]. The properties of single-phase ceramic materials do not meet all the requirements for these materials, therefore it is purposeful to use materials with a more complex structure. Ceramic matrix composites (CMCs) are systems designed to combine the excellent strength and high temperature properties of ceramics with the durability of advanced composites [3–5]. Ceramic composites in which the alumina matrix is reinforced with a wide range of ceramic phases including TiC, TiN, $ZrO_2$, WC, NbC, Ti(C,N), SiC, $TiB_2$, offer good strength properties, high hardness, wear resistance, chemical inertness and an improved toughness level in comparison to alumina [6–8]. The current interest in transition-metal carbides and nitrides is due to the many unique properties that these compounds exhibit. A large area of phase stability of many transition-metal carbides makes it possible to potentially tailor their thermomechanical and thermophysical properties. It is due to the crystal structure of carbides with a number of octahedral vacancies in the transition metal lattice. Titanium carbide (TiC) belongs to metal-like carbides also called interstitial carbides. Most of the metal-like carbides retain the structure of metal in which the carbon is located inside the octahedral vacancies. In titanium carbides, it is possible to incompletely fill the octahedral vacancies. Titanium forms only monocarbides $TiC_y$ with the widest homogeneity interval from $TiC_{1.0}$ to $TiC_{0.48}$ [9,10]. The combination of three types of bonds—covalent, ionic and metallic—in the structure of metal-like carbides causes the material to exhibit combined features of ceramic and metallic materials, showing both high hardness (28–35 GPa), high elastic modulus (450 GPa) as well as good thermal (22–35 W/(m·K)) and electrical conductivity [11]. Titanium carbide exhibits a high melting temperature (3340 K) and linear thermal expansion coefficient is $8.5 \times 10^{-6}$ $K^{-1}$ [12]. Both TiC and TiN crystallize in the structure of NaCl, where the corner of the face-centred-cubic (f.c.c.) unit cell is occupied by C or N atoms. Titanium nitride (TiN), exhibits excellent properties such as chemical stability, thermal stability, resistance to oxidation, high fracture toughness and high hardness (18–21 GPa). For the TiN phase, thermal conductivity (29.31 W/(m·K)) and Young modulus (466 GPa) are similar to that determined for TiC, whereas the thermal expansion coefficient is slightly higher at $9.4 \times 10^{-6}$ $K^{-1}$. The melting temperature of TiN is lower (3200 K) than that of TiC [13]. Different sintering technologies have been applied to produce the ceramic tool composites with improved mechanical properties. In pressureless sintering (PS), a low heating rate, high sintering temperature and long holding time are needed to obtain fully dense ceramic composites [14]. Spark plasma sintering (SPS) is widely used because of its great advantages compared to conventional sintering. High-speed powder consolidation, lower temperature of sintering, small holding time at the sintering temperature limit the precipitation of additional phases, resulting from unwanted reactions. SPS is a synthesis and processing technique which uses the power of a high electric pulse to generate plasma between powder particles in gaps or at the contact point, by cleaning powder surfaces (surface treatment) and increasing diffusivity at the physically contacted regions between particles [15]. The sample is heated by Joule-heating and the sparking among the particles of the sintered material leads instantaneously to faster heat and mass transfer. The basic configuration of a typical SPS system is shown in Figure 1a. Figure 1b illustrates the flow of pulsed current through powder particles inside an SPS sintering die [16].

The main objective of the present study was to help establish an effect of various spark plasma sintering temperatures (change in the range of about 100 °C) on the properties of TiC and TiN reinforced alumina–zirconia composites with different amounts of zirconia and compare with the effects of pressureless sintering. To satisfy the industrial requirements for precision machining of hard-working pieces, the results of this study could allow new titanium carbide and titanium nitride reinforced alumina–zirconia matrix composites to be obtained that combine toughness with increased hardness. The basic way to prepare alumina matrix composites reinforced with carbides and nitrides is to use a conventional pressureless (PS) method of sintering. Full densification of these composites requires a high temperature above 1700 °C. The direct electric discharge applied in the SPS process makes it possible to achieve a highly compacted bulk material at much lower temperature (by about 150 °C). So far, to the best of our knowledge, there have been no studies of the mechanical properties of $Al_2O_3/ZrO_2/(TiC + TiN)$ composites sintered by the SPS technique in a wide range of temperatures. Neither has there been any analysis of the influence of varying zirconium oxide content on the properties of these composites. The $ZrO_2$ whisker reinforced Ti(C,N) based cermets with addition of Co and WC were recently analysed [17]. Zirconia–alumina matrix composite reinforced with Ti(C,N) and Ni, Ti, Mo added as binders were sintered by microwave, repetitious-hot-pressing, and HIP (Hot

Isostatic Pressing) and tested [6,8,17–20]. Spark plasma sintering was used with good results for alumina ceramics and for Ti(C,N)-based cermets with the addition of a metal binder phase [21,22].

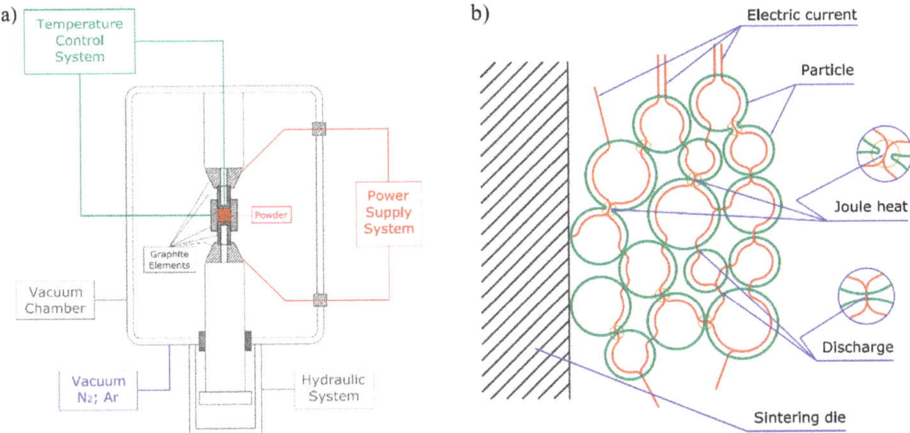


**Figure 1.** Spark plasma sintering (SPS) system: (**a**) basic configuration, (**b**) pulsed current flows through powder particles.

## 2. Materials and Methods

### 2.1. Materials

Commercially available powders of alumina, zirconia and titanium carbide as well as nitride were the starting materials. A high purity $\alpha$-$Al_2O_3$ (99.8%) powder type A16SG produced by Alcoa, United States with an average particle size of less than 0.5 μm and TiC, TiN powder in microscale (1–3 μm) produced by H.C. Starck, Germany were used. The monoclinic phase of zirconia in microscale $ZrO_2^{(m)}$ produced by Fluka, Germany and yttria-stabilized zirconia YSZ produced by H.C. Starck, Germany with a mean particle size of 1–3 μm were added in the amount of 2 wt%, 5 wt% for $ZrO_2^{(m)}$ and 10 wt% for $YZrO_2$. A small amount (less than 0.5 wt.%) of magnesium oxide was added to limit grain growth. The percentage weight fraction of various components are recorded in Table 1.

**Table 1.** Composition of compounds.

| Compound | Components, wt% | | | | |
|---|---|---|---|---|---|
| | $Al_2O_3$ + MgO | $ZrO_2^{(m)}$ | $YZrO_2$ | TiC | TiN |
| TCN1 | 68 | 2 | - | 10 | 20 |
| TCN2 | 65 | 5 | - | 10 | 20 |
| TCN3 | 60 | - | 10 | 10 | 20 |

Components with plasticiser were mixed in alumina mills with zirconia balls, for thirty hours. The powders after plasticising and drying, were granulated and directly fed into the graphite dies. The compounds were consolidated in a spark plasma sintering system at different temperatures from 1575 °C to 1675 °C at a heating rate of 100 °C/min and under a pressure of 35 MPa. Holding time at peak temperature was 10 min. Figure 2 presents the change in the temperature and in piston displacement during SPS sintering of alumina–zirconia powders with TiC and TiN.

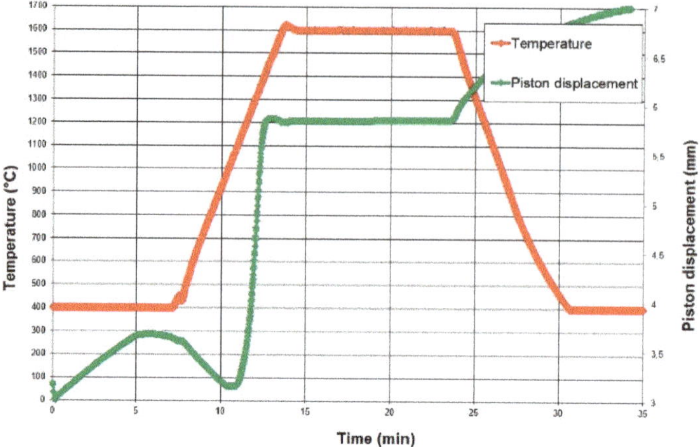

**Figure 2.** Sintering curve of alumina–zirconia powders with TiC and TiN.

Furnace type FCT HD5 (FCT Systeme GmbH, Frankenblick, Germany) under argon conditions was used. The final sintered sample was in a shape of a disk with a diameter of 20 mm and about 5.0 mm in thickness. Furthermore, the samples were sintered by pressureless sintering (PS) in the HTK8 Gero furnace (Carbolite Gero GmbH & Co. KG, Neuhausen, Germany) in argon atmosphere, at a 1750 °C temperature and a heating rate of 100 °C/min. The temperature was controlled using an optical pyrometer. The sintering time was 1 h. After sintering the specimens were plane ground with diamond (MD-Piano120), and also fine ground with diamond grains of about 3 μm (MD-Largo). Before measuring the mechanical properties and microstructural observation the specimens were polished with an OP-S type colloidal silica (size of silicon oxide grains of about 0.04 μm). The roughness $R_a$ of the tested surface was not more than 0.1 μm.

### 2.2. Characterisation

The following measurements were performed: Vickers hardness HV1, Young's modulus $E$, friction coefficient μ, (in ball-on-disc tests, using a UMT-2MT universal mechanical tester produced by CETR, Campbell, CA, USA), relative density and apparent density $\rho$, according to the EN 623-2 standard [23]. The Vickers hardness HV1 was measured under a load of 9.81 N using a digital hardness tester (FV-700 by FUTURE-TECH CORP., Kawasaki, Japan). The standard deviations of five hardness measurements made for each sample did not exceed 2% of the average values. Ultrasonic method based on velocity measurements of transversal and longitudinal waves in the samples has been used to determine the Young's modulus E. The test stand for ultrasonic measurements consisted of an ultrasonic flaw detector (Epoch III by Panametrics, Waltham, MA, USA), a set of special broadband heads emitting longitudinal or transversal waves. The whole system operated under the control of a specialized software for data collection and processing. The following formula was used for calculation of Young's modulus [24]:

$$E = \rho C_T^2 \frac{3C_L^2 - 4C_T^2}{C_L^2 - C_T^2} \tag{1}$$

where $\rho$—apparent density, $C_L$—velocity of the longitudinal wave, $C_T$—velocity of the transversal wave.

The point indentation technique has found application in the study of the indentation fracture toughness. The indentation fracture toughness $K_{IC(HV)}$ was evaluated from a direct measurement of the crack length as a function of the indentation load called the "direct crack measurement" (DCM). The "halfpenny" shaped, elliptical cracks formed during the Vickers hardness indentation gave a basis to determine the indentation fracture toughness $K_{IC(HV)}$. For these cracks the ratio c/a $\geq$ 2.5 was observed and the indentation fracture toughness was defined as follows [25]:

$$(K_{IC(HV)}\ \varphi/H\ a^{1/2})\ (H/E\ \varphi)^{2/5} = 0.129\ (c/a)^{-3/2} \tag{2}$$

where $K_{IC(HV)}$ is the critical stress intensity factor, $\varphi$ is the constrain factor, $H$ is the Vickers hardness, $E$ is the Young's modulus, a is half the length of indent diagonal, c is the sum of a and l (crack length).

The pressing-in load used to produce the Vickers cracks was of 294.3 N. Tribological research using the ball-on-disc method, with the use of a universal UMT-2-MT mechanical equipment was carried out. The friction coefficient and the specific wear rate of the disk-shaped samples in contact with $Al_2O_3$ ball were measured. In this method, sliding contact is possible by pressure of the ball specimen onto a rotating disk specimen under a constant load. Using a controlled load $F_n$ to the ball holder, the friction force was measured continuously during the test with an extensometer. For each test, a new ball was used. The ball and disk samples were washed in ethyl alcohol and dried. The diameter of the disk-shaped samples was of 20 mm and about 5.0 mm in height. The following test conditions were adopted: ball diameter of 3.2 mm, applied load of 20 N, sliding speed of 0.1 m/s, radius of the sliding circle of 4 mm, sliding distance of 2000 m and time of the test of 20,000 s. The specimens were tested without lubricant at room temperature. Each test was repeated on three specimens. The friction coefficient was expressed as follows:

$$\mu = \frac{F_f}{F_n} \tag{3}$$

where $\mu$—friction coefficient, $F_f$—measured friction force, $F_n$—applied normal force.

After the test, the cross-sectional profile of the wear track in four places at intervals of 90° was measured using a contact stylus profilometer PRO500 (CETR, Campbell, CA, USA), according to the ISO 20808:2016 standard [26]. Then, the average cross-sectional area of the wear track was calculated. The volume of removed material was evaluated as a product of the cross-sectional area of the wear track and its circumference. The specific wear rate can then be defined using Equation (4):

$$Ws = \frac{V}{F_n \cdot L} \tag{4}$$

where $W_s$—specific wear rate, $V$—volume of removed material, $L$—sliding distance.

The microstructure observations of the specimen were characterised using JEOL JSM-6460LV (JEOL Ltd., Tokyo, Japan) scanning electron microscope equipped with EDS (Energy-dispersive X-ray spectroscopy), EBSD (Electron Backscatter Diffraction), WDS (Wavelength-Dispersive X-ray Spectroscopy). XRD measurements were taken using an Empyrean system (PANAlytical, Malvern, UK) with Cu Kα1 radiation. The phase compositions of the sintered compacts were identified using the database of International Centre for Diffraction Data (Newtown Square, PA, USA) PDF4+2018. The quantitative compositions of the tested composites were determined by the Rietveld method with the X'Pert Plus program (PANAlytical, Malvern, UK).

## 3. Results and Discussion

Table 2 shows the average values of the following properties of the tested alumina matrix composites TCN1, TCN2, TCN3 at sintering temperatures from 1575 °C to 1675 °C: apparent density, relative density and Young modulus. Samples of the TCN1 composition sintered by pressureless sintering (PS) were marked TCN1*.

**Table 2.** Apparent density, relative density and Young modulus of the tested composites.

| Sample | Sintering Temperature (°C) | Apparent Density $\rho$ (g/cm$^3$) | Relative Density (%) | Young Modulus $E$ (GPa) |
|---|---|---|---|---|
| TCN1.1 | 1575 | 4.26 ± 0.01 | 98.18 | 412 ± 5.0 |
| TCN1.2 | 1600 | 4.29 ± 0.01 | 98.87 | 418 ± 5.0 |
| TCN1.3 | 1625 | 4.28 ± 0.01 | 98.64 | 416 ± 5.0 |
| TCN1.4 | 1650 | 4.29 ± 0.01 | 98.87 | 415 ± 4.5 |
| TCN1.5 | 1675 | 4.29 ± 0.01 | 98.87 | 421 ± 4.5 |
| TCN2.1 | 1575 | 4.34 ± 0.01 | 98.89 | 411 ± 5.0 |
| TCN2.2 | 1600 | 4.33 ± 0.01 | 98.67 | 405 ± 5.0 |
| TCN2.3 | 1625 | 4.34 ± 0.01 | 98.89 | 412 ± 5.0 |
| TCN2.4 | 1650 | 4.34 ± 0.01 | 98.89 | 407 ± 4.5 |
| TCN2.5 | 1675 | 4.33 ± 0.01 | 98.67 | 406 ± 5.0 |
| TCN3.1 | 1575 | 4.43 ± 0.01 | 98.01 | 404 ± 4.5 |
| TCN3.2 | 1600 | 4.43 ± 0.01 | 98.01 | 400 ± 5.0 |
| TCN3.3 | 1625 | 4.44 ± 0.01 | 98.23 | 403 ± 4.5 |
| TCN3.4 | 1650 | 4.44 ± 0.01 | 98.23 | 407 ± 5.0 |
| TCN3.5 | 1675 | 4.44 ± 0.01 | 98.23 | 405 ± 5.0 |
| TCN1* | 1750 | 4.21 ± 0.01 | 97.03 | 378 ± 4.0 |

Notice: TCN1* specimen was sintered by the pressureless sintering (PS) method.

The values of Vickers hardness *HV1* and indentation fracture toughness $K_{IC(HV)}$ are presented in Figures 3 and 4.

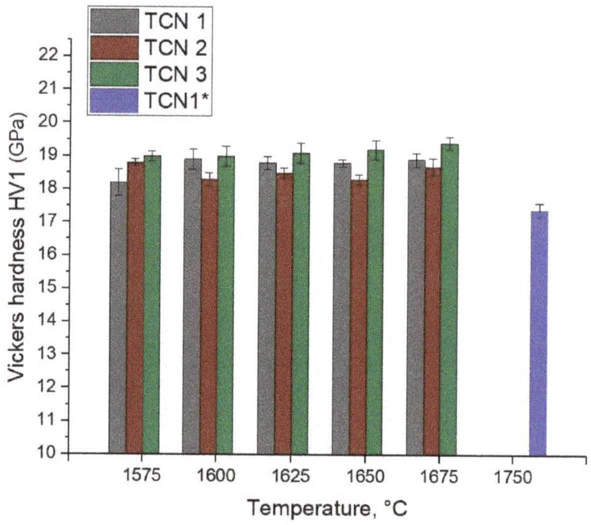


**Figure 3.** Vickers hardness versus sintering temperature for the tested composites: TCN1, TCN2, TCN3 and TCN1*.

Titanium carbide and nitride reinforced alumina matrix composites consolidated through SPS exhibited: Vickers hardness *HV1* from 19.4 GPa to 18.2 GPa, critical stress intensity factor $K_{IC(HV)}$ from 3.84 MPa·m$^{1/2}$ to 5.83 MPa·m$^{1/2}$, Young modulus from 400 GPa to 421 GPa, apparent density from 4.26 g/cm$^3$ to 4.44 g/cm$^3$, and relative density from 98.18% to 98.89%. Samples sintered by SPS marked as TCN3 had the Vickers hardness value of over 19.0 GPa. The results showed no significant influence of the sintering temperature on the mechanical properties of these composites. There was only 1–2% difference in Vickers hardness between TCN3 samples measured at the lowest (1575 °C) and highest (1675 °C) temperatures. The Vickers hardness for TCN1 and TCN2 composites was slightly lower and was in the range between 18.2 and 18.9 GPa. The effect of the sintering temperature on the Vickers hardness HV1 was not observed for TCN1 and TCN2 composites manufactured with the content of the $ZrO_2^{(m)}$ zirconia phase of 2 wt% and 5 wt% respectively. However, an effect of the sintering temperature on the indentation fracture toughness of the tested composites was observed (Figure 2). The indentation fracture toughness of the tested composites increased by 10–20% as the temperature used was increased. TCN1 and TCN2 composites show a high value of $K_{IC(HV)}$ (about 6.0 MPa·m$^{1/2}$) at the highest temperature of 1675 °C. $K_{IC(HV)}$ of TCN3 at 1675 °C was about 4.4 MPa·m$^{1/2}$. After pressureless sintering of the TCN1* specimen at the highest temperature of 1750 °C, the following mechanical and physical properties were obtained: Vickers hardness *HV1* of 17.4 GPa, critical stress intensity factor $K_{IC(HV)}$ of 4.46 MPa·m$^{1/2}$, Young modulus of 378 GPa, apparent density of 4.21 g/cm$^3$, and relative density of 97.03%. Although such a high sintering temperature was used for this specimen, its mechanical properties were 10–25% lower compared to specimens sintered at the highest temperature (1675 °C) and even 5–10% lower compared to those sintered at the lowest temperature (1575 °C). The X-ray diffraction analysis of the composites with a different amount of zirconia phase at the sintering temperature of 1600 °C (marked as TCN1.2, TCN2.2 and TCN3.2) indicated the presence of the following phases: $Al_2O_3$, Ti(C,N), TiC, TiN, $ZrO_2$ in monoclinic and tetragonal forms and titanium aluminium oxide ($Ti_{0.25}Al_{1.75}O_3$)—see Figure 5.

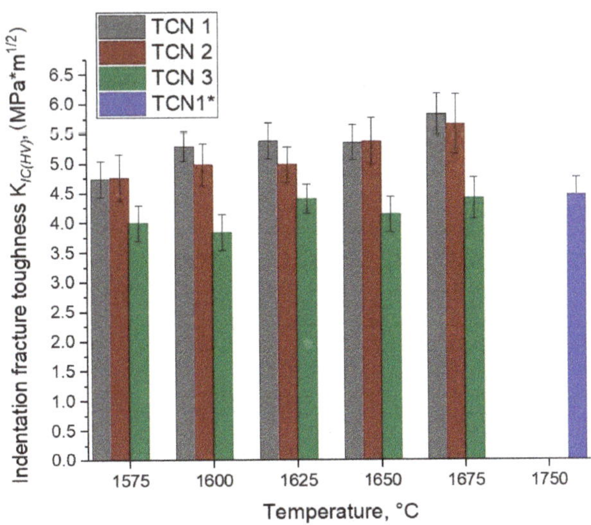


**Figure 4.** Indentation fracture toughness versus sintering temperature for the tested composites: TCN1, TCN2, TCN3 and TCN1*.

The peak of the titanium aluminium oxide was not visible in the X-ray diffraction pattern because it correlated with the alumina peak. The quantitative phase composition obtained by X-ray diffraction analysis of the composites with 2 wt% of $ZrO_2^{(m)}$, 5 wt% of $ZrO_2^{(m)}$, and 10 wt% of $YZrO_2^{(t)}$ (marked as TCN1, TCN2, TCN3, respectively) is presented in Table 3.

**Table 3.** X-ray diffraction analysis of the tested composites.

| Composite | Content (wt%) | | | | | |
|---|---|---|---|---|---|---|
| | $Al_2O_3$ | TiN | TiC | Ti(C,N) | $ZrO_2^{(m)}$ | $ZrO_2^{(t)}$ |
| TCN1 | 68.0 | 17.0 | 8.0 | 5.0 | 1.0 | 1.0 |
| TCN2 | 66.0 | 14.0 | 4.0 | 12.0 | 2.0 | 2.0 |
| TCN3 | 59.0 | 12.0 | 4.0 | 15.0 | - | 10.0 |

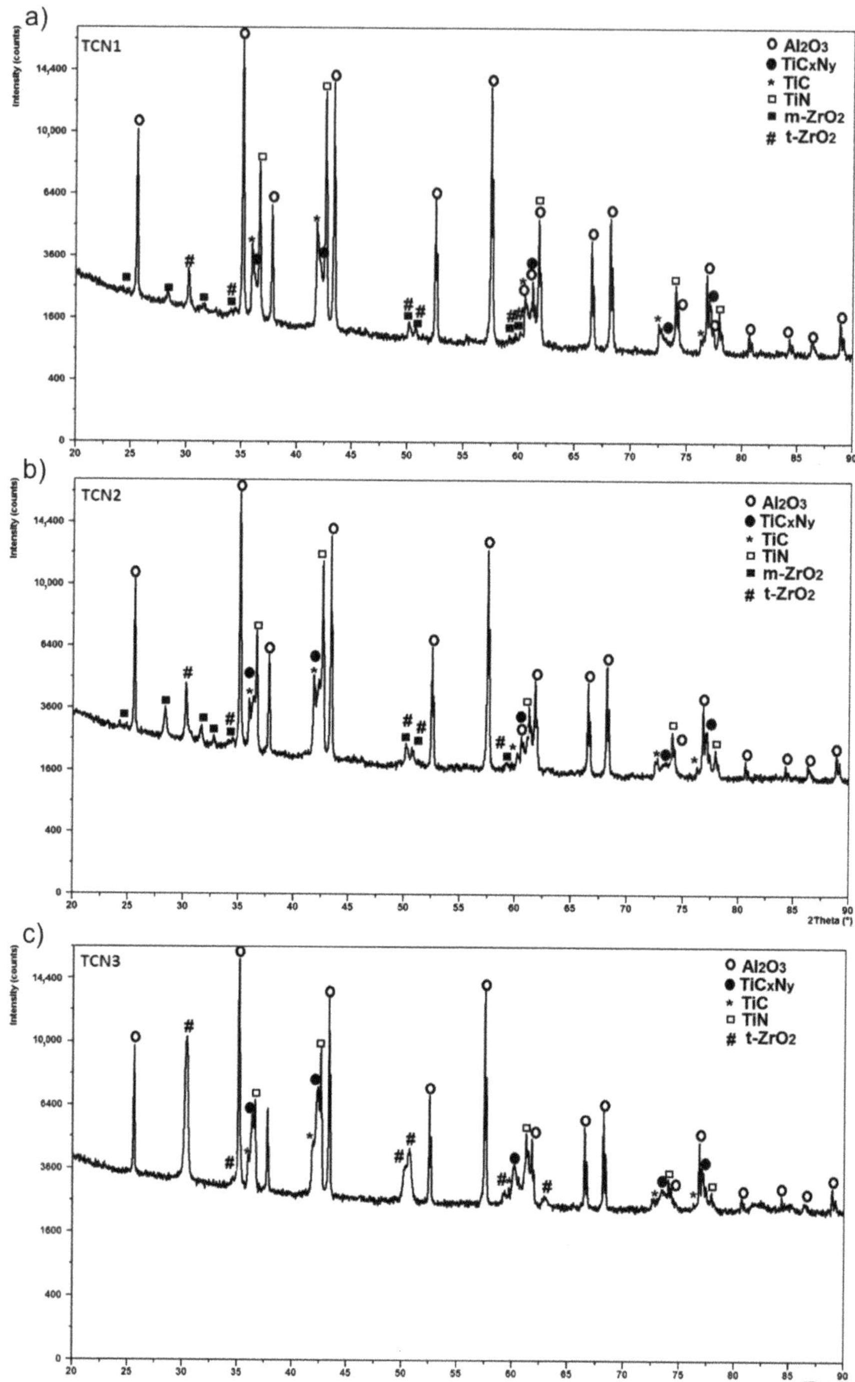

**Figure 5.** X-ray diffraction pattern of the $Al_2O_3/ZrO_2/(TiC + TiN)$ composites: (**a**) TCN1; (**b**) TCN2; (**c**) TCN3.

Tetragonal $ZrO_2^{(t)}$ and monoclinic $ZrO_2^{(m)}$ phases in the specimens TCN1 and TCN2 with the addition of 2 wt% and 5 wt% monoclinic $ZrO_2^{(m)}$ were observed. This was due to partial polymorphic transformation with the martensitic nature of monoclinic $ZrO_2^{(m)}$ into tetragonal $ZrO_2^{(t)}$. The TCN3 composite revealed only zirconia tetragonal phase $ZrO_2^{(t)}$. The analysis of the X-ray diffraction of the tested composites showed the highest amount of Ti(C,N) was a continuous solution of TiN and TiC for the TCN3 specimen. A SEM image of the indentation crack path in $Al_2O_3/ZrO_2/(TiC + TiN)$ composites (marked as TCN1 and TCN3) is presented in Figure 6a,b.

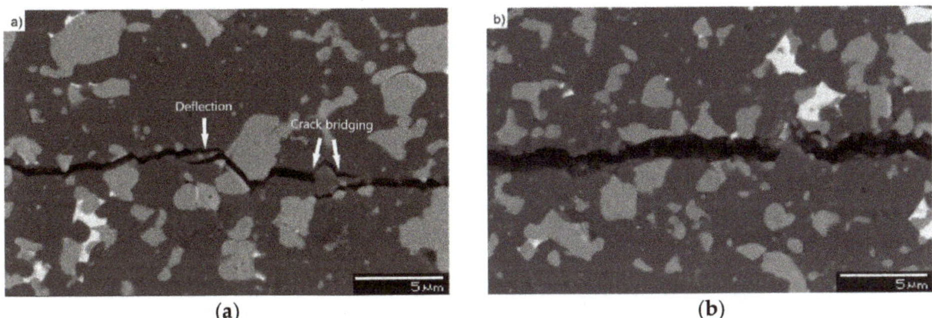


**Figure 6.** SEM image of indentation crack path in $Al_2O_3/ZrO_2/(TiC + TiN)$ composites:(**a**) TCN1.5; (**b**) TCN3.5.

The microstructures of the tested composites revealed transgranular and intergranular fractures during crack propagation. The presence of the transgranular fracture is favourable for the enhancement of toughness in these composites. Furthermore, crack bridging and crack deflection were observed for TCN1 specimen (Figure 6a). The transgranular fracture absorbs more energy, and it improves the strength properties of composites. The presence of the crack bridging can provide a force to make the two surfaces of the crack draw closely, thus the crack propagation is limited [14].

SEM microstructural observations of the ceramic composites showed their good consolidation. The EDS technique was used to carry out surface distribution analysis and semi-quantitative microanalysis of the elements present in the specimens, revealing the following phases: TiN, TiC, Ti(C,N), $ZrO_2$, MgO (Figure 7).

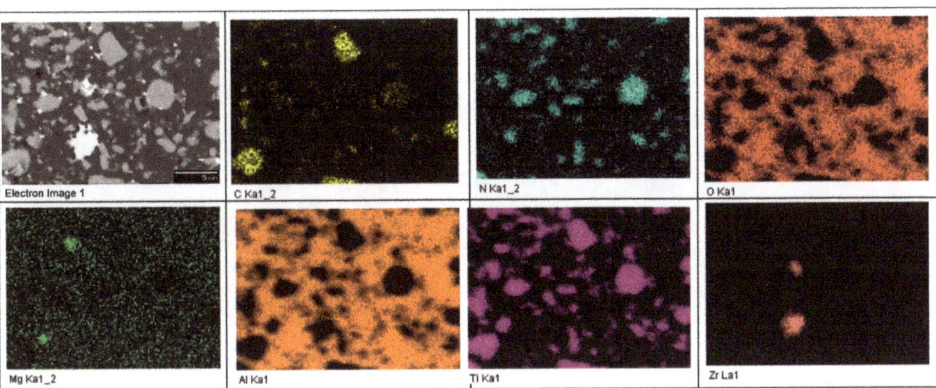


**Figure 7.** Distribution of the elements in the TCN1.1 composite.

Figure 8 presents microstructure images for the tested $Al_2O_3/ZrO_2/(TiC + TiN)$ composites sintered at 1575 °C and 1675 °C. The microstructures of the composites with an addition of 2 wt% and 5 wt%

$ZrO_2^{(m)}$ (marked TCN1 and TCN2, respectively) sintered at 1575 °C were similar. TiC and TiN grains of differential sizes (fine to coarse) and shapes were distributed in the alumina matrix. Apart from single zirconia grains, zirconia grain agglomerates also occurred (Figure 8a,c). The average grain size of TiC and TiN is within the range of 2.7 µm to 3.0 µm. In TCN3 composites with the addition of 10 wt% yttria-stabilised zirconia, TiC and TiN grains exhibit a similar shape and a smaller size in comparison with TCN1 and TCN2 composites. The average grain size of TiC and TiN was about 2.4 µm. However, the grains of Ti(C,N) solid solution in the composites sintered at 1675 °C were observed to have a rounder shape (Figure 8b,d,f). The grain size of Ti(C,N) was from 1.9 µm to 2.3 µm. The presence of Ti(C,N) solid solution in the microstructure of the composites changed their properties. The friction coefficient for the tested composites was in the range of 0.31 to 0.55 (Figure 9).

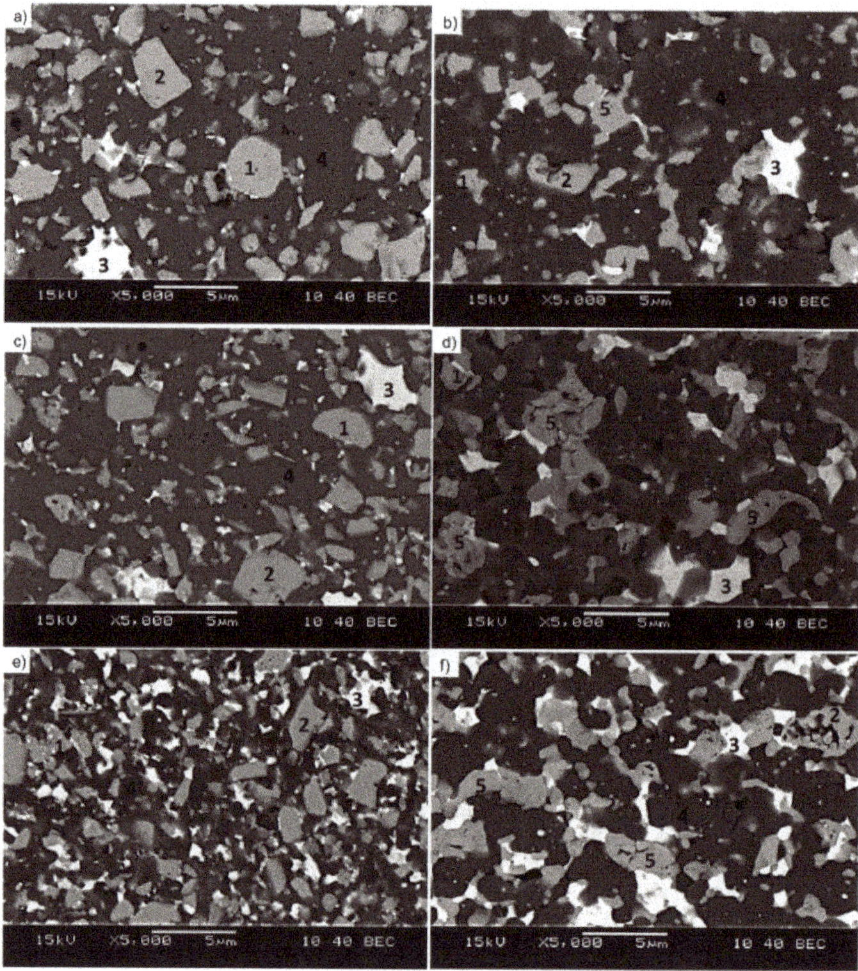

**Figure 8.** SEM microstructure of the $Al_2O_3/ZrO_2/(TiC + TiN)$ composites for: (**a**) TCN1.1, (**b**) TCN1.5, (**c**) TCN2.1, (**d**) TCN2.5, (**e**) TCN3.1, (**f**) TCN3.5, where: 1—TiN, 2—TiC, 3—$ZrO_2$, 4—$Al_2O_3$, 5—Ti(C,N).

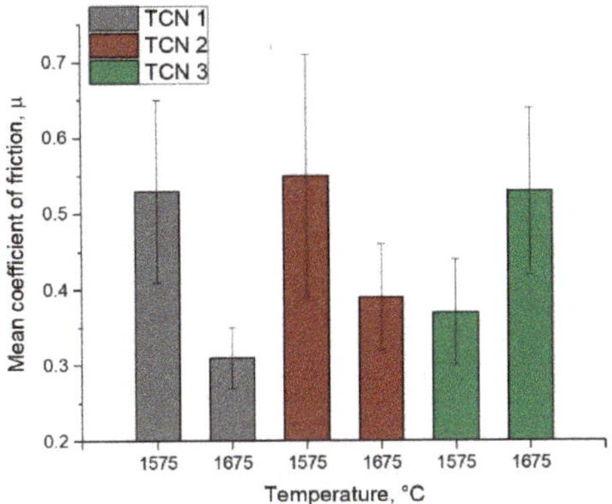

**Figure 9.** Mean friction coefficient of the tested composites.

An increased sintering temperature of up to 1675 °C resulted in a lower friction coefficient for specimens TCN1 and TCN2 with the addition of 2 wt% and 5 wt% monoclinic $ZrO_2$ phase, respectively, whereas the TCN3 composite with 10 wt% of yttria-stabilised zirconia (YSZ) exhibited an inverse correlation.

For all the composites tested, a higher sintering temperature caused an increase in the wear rate of both the disk and the ball (Figure 10).

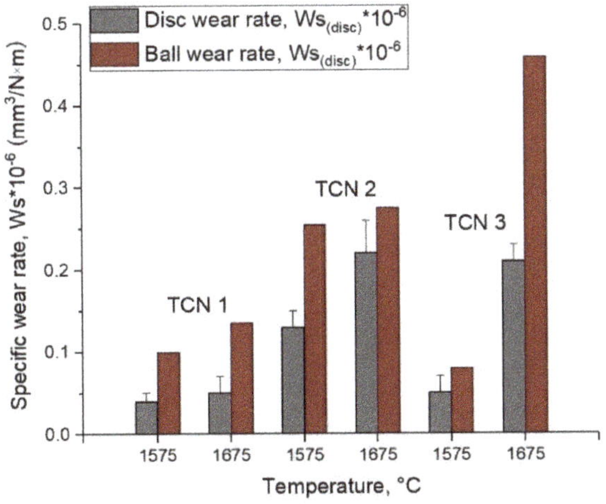

**Figure 10.** Specific wear rate for disc and ball for tested composites.

The improvement of mechanical properties were attributed to the increased rate of densification in spark plasma sintering and to microstructural changes. The presence of Ti(C,N) solid solution in the microstructure of the composites changed their properties. The content of 15 wt% Ti(C,N) solid solution with high Vickers hardness of 30.0 GPa, with the TiC phase reduced to 40 wt.% resulted in a higher Vickers hardness and lower indentation fracture toughness of the specimens. Moreover, character of fracture and crack propagation was an additional parameter in improvement properties. Both the transgranular fracture and the crack bridging observed for crack propagation resulted in improved fracture toughness of tested alumina–zirconia matrix composites. Using the SPS technique to consolidate alumina–zirconia powder with the addition of TiC and TiN, it was possible to obtain a high-density ceramic material with the following characteristics: over 98% density in the temperature range of 1575–1675 °C, Vickers hardness of about 19.0 GPa, fracture toughness of over 5.0 MPa m$^{1/2}$. These properties were significantly better than those obtained by pressureless sintering. To the best of our knowledge, no study results of TiC- and TiN-reinforced alumina–zirconia composites (with various amount of zirconia) manufactured by SPS technique have been presented in literature. Therefore, it is not possible to compare the results obtained in this study with other results obtained in similar conditions. While the influence of the most important operating parameters on product characteristics after consolidation/synthesis of ceramic materials by electric current activated/assisted was widely discussed in the last decade [27,28], attention was focussed on alumina ceramics, alumina–zirconia composites and alumina matrix composites reinforced with SiC, TiC, Ti$_3$SiC$_2$, TiN, graphene platelets (GPls) [29]. Al$_2$O$_3$ ceramics were super-fast densified by SPS at temperatures ranging between 1350 °C and 1700 °C (600 °C/min heating rate) at 40 MPa, without selecting a holding time. Full densification of the samples was observed at sintered temperatures below 1550 °C. However, full densification of Al$_2$O$_3$–40 wt% TiC nanocomposite powders was achieved at a sintering temperature of 1480 °C (heating rate 50 °C/min) for 4 min and pressure 50 MPa by SPS technique. For these parameters of sintering a high value of the Vickers hardness equal to 21.0 GPa and low fracture toughness of 3.87 MPa m$^{1/2}$ were obtained. Al$_2$O$_3$–TiC composite powders synthesized by high-energy ball milling were consolidated near to theoretical density (99.6%) in SPS at a sintering temperature of 1450 °C for 4 min at 50 MPa [30]. Grain sizes of both Al$_2$O$_3$ and TiC were less than 1 mm. The Vickers hardness of 19.1 GPa and fracture toughness of 4.5 MPa m$^{1/2}$ were obtained. Al$_2$O$_3$/TiN nanocomposites with cubic TiN (46 vol%) and α-Al$_2$O$_3$ (54 vol%) were synthesized by in situ reactive synthesis (RS) (SPSRS—spark plasma sintering reactive synthesis) at a temperature of 1400 °C, under a pressure of 60 MPa, at different dwelling times (0–6 min) starting from a mixture of TiO$_2$, AlN and Ti powders [31]. Fine microstructure with TiN particles homogeneously distributed in the Al$_2$O$_3$ matrix were observed. The Vickers hardness and fracture toughness of samples from an SPSRS were 17.8 GPa, 4.22 MPa m$^{1/2}$ respectively. For comparison, Al$_2$O$_3$/TiN composites were also prepared at a temperature of 1420 °C under 60 MPa, using mechanically mixed TiN and Al$_2$O$_3$. These composites revealed slightly higher hardness and fracture toughness.

## 4. Conclusions

In this work, effort has been made to produce alumina–zirconia matrix composites reinforced with TiC and TiN phases as a tool material with improved mechanical properties. This was achieved by using SPS to consolidate the alumina–zirconia powder with the addition of TiC and TiN.

A high-density alumina–zirconia matrix composites with over 98% density in the temperature range of 1575–1675 °C, Vickers hardness of about 19.0 GPa, fracture toughness of over 5.0 MPa m$^{1/2}$ were manufactured. These properties were significantly better than those obtained by pressureless sintering at the highest temperature of 1750 °C.

The mechanical properties of the samples sintered at the temperature of 1750 °C by PS exhibited 5–10% lower values when compared to those sintered at the lowest temperature (1575 °C) by the SPS method. A lower relative density of 97% was achieved for the pressureless sintering at a temperature of 1750 °C.

No influence of the sintering temperature on the Vickers hardness of these composites was observed. SPS samples achieved comparable Vickers hardness and Young modulus in the range of the tested temperatures. However, slightly increase of the fracture toughness at the highest temperature was attained. An increase in the wear rate of both the disk and the ball of the tested composites at the higher sintering temperature was observed.

There are differences in mechanical properties of samples with various compositions of zirconia. Samples with tetragonal, yttria-stabilised zirconia (YSZ) phase sintered by SPS, marked as TCN3, had the highest Vickers hardness value of over 19.0 GPa. Samples with monoclinic zirconia phase of 2 wt% and 5 wt% sintered by SPS, marked as TCN1 and TCN2 respectively, had the Vickers hardness value less than 5% lower than that of TCN3 samples. This was related to the highest amount of Ti(C,N) as a continuous solution of TiN and TiC for the samples with tetragonal zirconia specimen shown in X-ray diffraction and microstructural observation.

A 10–30% increase of the fracture toughness was observed for samples with monoclinic zirconia phase when compared to samples with tetragonal zirconia phase.

**Author Contributions:** Conceptualization, M.S.; methodology, M.S.; validation, M.S.; formal analysis, M.S.; investigation, M.S., S.C., M.P., J.L.-Ł., J.C. and A.K.; resources, M.S.; data curation, M.S.; writing—original draft preparation, M.S.; writing—review and editing, M.S.; visualization, S.C.; supervision, M.S.; project administration, M.S.; funding acquisition, M.S.

**Funding:** This research and APC was funded by the National Science Centre Poland, grant number UMO-2016/21/B/ST8/01027.

**Acknowledgments:** This work was supported by the National Science Centre, Poland (project UMO-2016/21/B/ST8/01027) 2017–2020.

**Conflicts of Interest:** The authors declare no conflict of interest.

## References


1. Bobzin, K. High-performance coatings for cutting tools. *CIRP J. Manuf. Sci. Technol.* **2017**, *18*, 1–9. [CrossRef]
2. Pampuch, R. Kompozyty ceramiczne. *Kompozyty* **2002**, *2*, 10–14.
3. Tiwari, A.; Gerhardt, R.A.; Szutkowska, M. *Advanced Ceramic Materials*; John Wiley & Sons: Hoboken, NJ, USA, 2016; ISBN 1119242738.
4. Szutkowska, M.; Boniecki, M. Fracture toughness of alumina matrix composites (AMC). *J. Optoelectron. Adv. Mater.* **2010**, *12*, 301–308.
5. Ni, X.; Zhao, J.; Sun, J.; Gong, F.; Li, Z. Effects of metal binder on the microstructure and mechanical properties of $Al_2O_3$-based micro-nanocomposite ceramic tool material. *Int. J. Miner. Metall. Mater.* **2017**, *24*, 826–832. [CrossRef]
6. Yin, Z.; Huang, C.; Zou, B.; Liu, H.; Zhu, H.; Wang, J. Preparation and characterization of $Al_2O_3$/TiC micro–nano-composite ceramic tool materials. *Ceram. Int.* **2013**, *39*, 4253–4262. [CrossRef]
7. Acchar, W.; Cairo, C.A. The influence of (Ti, W) C and NbC on the mechanical behavior of alumina. *Mater. Res.* **2006**, *9*, 171–174. [CrossRef]
8. Song, S.-X.; Ai, X.; Zhao, J.; Huang, C.-Z. $Al_2O_3$/Ti ($C_{0.3}N_{0.7}$) cutting tool material. *Mater. Sci. Eng. A* **2003**, *356*, 43–47. [CrossRef]
9. Rempel, A.A. Nonstoichiometric transition metal compounds. A review. In Proceedings of the 17th Israeli-Russian Bi-National Workshop 2018, the Optimization of the Composition, Structure and Properties of Metals, Oxides, Composites, Nano and Amorphous Materials, Moscow, Russia, 13–17 August 2018; pp. 167–189.
10. Gusev, A.I.; Rempel, A.A.; Magerl, A.J. *Disorder and Order in Strongly Nonstoichiometric Compounds: Transition Metal Carbides, Nitrides and Oxides*; Springer Science & Business Media: Berlin/Heidelberg, Germany, 2013; Volume 47, ISBN 3662045826.
11. Lengauer, W. Transition metal carbides, nitrides, and carbonitrides. *Handb. Ceram. Hard Mater.* **2000**, 202–252. [CrossRef]
12. Rajabi, A.; Ghazali, M.J.; Daud, A.R. Chemical composition, microstructure and sintering temperature modifications on mechanical properties of TiC-based cermet—A review. *Mater. Des.* **2015**, *67*, 95–106. [CrossRef]
13. Yu, S.; Zeng, Q.; Oganov, A.R.; Frapper, G.; Zhang, L. Phase stability, chemical bonding and mechanical properties of titanium nitrides: A first-principles study. *Phys. Chem. Chem. Phys.* **2015**, *17*, 11763–11769. [CrossRef]
14. Yin, Z.; Yuan, J.; Wang, Z.; Hu, H.; Cheng, Y.; Hu, X. Preparation and properties of an $Al_2O_3$/Ti (C, N) micro-nano-composite ceramic tool material by microwave sintering. *Ceram. Int.* **2016**, *42*, 4099–4106. [CrossRef]
15. Rozmus, M.; Putyra, P.; Figiel, P. The influence of non-conventional sintering methods on grain growth and properties of alumina sinters. *Mechanik* **2015**, *88*, 249–258. [CrossRef]

16.	Tokita, M. Trends in Advanced SPS Spark Plasma Sintering Systems and Technology. Functionally Gradient Materials and Unique Synthetic Processing Methods from Next Generation of Powder Technology. *J. Soc. Powder Technol. Japan* **1993**, *30*, 790–804. [CrossRef]

17.	Fang, Y.; Zhao, X.; Zhang, M.; Zhang, P.; Cheng, H.; Xue, S.; Zhang, L.; Wu, J.; Feng, S. Effect of $ZrO_2$ whiskers on the microstructure and mechanical properties of a Ti (C, N)-based cermet cutting tool material. *Int. J. Appl. Ceram. Technol.* **2019**. [CrossRef]

18.	Li, Q.; Sun, X.D.; Xiu, Z.M. Microstructure and properties of $Al_2O_3$/TiCN–Ni–Ti composite fabricated by hot pressing. *Chin. J. Nonferrous Met.* **2012**, *22*, 2311–2316.

19.	Xikun, L.; Guanming, Q.; Tai, Q.; Haitao, Z.; Hua, B.; Xudong, S. $Al_2O_3$/TiCN-0.2% $Y_2O_3$ composite prepared by HP and its cutting performance. *J. Rare Earths* **2007**, *25*, 37–41. [CrossRef]

20.	Dyatlova, Y.G.; Agafonov, S.V.; Boikov, S.Y.; Ordanyan, S.S.; Rumyantsev, V.I. Tool ceramics based on $Al_2O_3$–$ZrO_2$–TiCN composite. *Powder Metall. Met. Ceram.* **2011**, *49*, 675–681. [CrossRef]

21.	Shen, Z.; Johnsson, M.; Zhao, Z.; Nygren, M. Spark plasma sintering of alumina. *J. Am. Ceram. Soc.* **2002**, *85*, 1921–1927. [CrossRef]

22.	Zheng, Y.; Wang, S.; You, M.; Tan, H.; Xiong, W. Fabrication of nanocomposite Ti (C, N)-based cermet by spark plasma sintering. *Mater. Chem. Phys.* **2005**, *92*, 64–70. [CrossRef]

23.	British Standards Institution. *European Standard EN 623-2:1993 Advaced Technical Ceramics—Monolithic Ceramics—General and Textural Properties—Part 2—Determination of Density and Porosity*; British Standards Institution: London, UK, 1993.

24.	Klimczyk, P. SiC-Based Composites Sintered with High Pressure Method. *Silicon Carbide Mater. Process. Appl. Electron. Devices* **2011**, 309–334. [CrossRef]

25.	Niihara, K.; Morena, R.; Hasselman, D.P. Evaluation of Klc of Brittle Solids By the Indentation Method With Low Crack To Indent Ratios. *J. Mater. Sci. Lett.* **1982**, *1*, 13–16. [CrossRef]

26.	ISO INTERNATIONAL STANDARD. *Fine Ceramics (Advanced Ceramics, Advanced Technical Ceramics)—Determination of Friction and Wear Characteristics of Monolithic Ceramics by Ball-on-Disc Method*; ISO: Geneva, Switzerland, 2016.

27.	Liu, J.; Li, Z.; Yan, H.; Jiang, K. Spark plasma sintering of alumina composites with graphene platelets and silicon carbide nanoparticles. *Adv. Eng. Mater.* **2014**, *16*, 1111–1118. [CrossRef]

28.	Tokita, M.; Tamari, N.; Takeuchi, T.; Makino, Y. Consolidation behavior and mechanical properties of SiC with $Al_2O_3$ and $Yb_2O_3$ consolidated by SPS. *J. Jpn. Soc. Powder Powder Metall.* **2009**, *56*, 788–795. [CrossRef]

29.	Orru, R.; Licheri, R.; Locci, A.M.; Cincotti, A.; Cao, G. Consolidation/synthesis of materials by electric current activated/assisted sintering. *Mater. Sci. Eng. R Rep.* **2009**, *63*, 127–287. [CrossRef]

30.	Zhang, Y.F.; Wang, L.J.; Jiang, W.; Chen, L.D. Preparation of $Al_2O_3$-TiC Composites by High Energy Ball Milling Combined with SPS Technology and Its Properties. *J. Inorg. Mater.* **2005**, *20*, 1445–1449.

31.	Wang, L.; Wu, T.; Jiang, W.; Li, J.; Chen, L. Novel fabrication route to $Al_2O_3$–TiN nanocomposites via spark plasma sintering. *J. Am. Ceram. Soc.* **2006**, *89*, 1540–1543. [CrossRef]

*Article*

# Microstructure, Microhardness, and Wear Properties of Cobalt Alloy Electrodes Coated with TiO$_2$ Nanoparticles


**Sebastian Balos [1,*], Petar Janjatovic [1,*], Miroslav Dramicanin [1], Danka Labus Zlatanovic [1], Branka Pilic [2], Pavel Hanus [3] and Lucyna Jaworska [4]**

[1]   Faculty of Technical Sciences, University of Novi Sad, 21000 Novi Sad, Serbia; dramicanin@uns.ac.rs (M.D.); danlabus@uns.ac.rs (D.L.Z.)

[2]   Faculty of Technology, University of Novi Sad, 21000 Novi Sad, Serbia; brapi@uns.ac.rs

[3]   Department of Material Science, Technical University of Liberec, Liberec 461 17, Czech Republic; Pavel.Hanus@tul.cz

[4]   Lukasiewitz- Research Network-Institute of Advanced Manufacturing Technology, Centre for Materials Research and Sintering Technology, 30-011 Krakow, Poland; lucyna.jaworska@ios.krakow.pl

*   Correspondence: sebab@uns.ac.rs (S.B.); janjatovic@uns.ac.rs (P.J.); Tel.: +381-21-485-2339 (S.B.); +381-21-485-2323 (P.J.)


Received: 27 September 2019; Accepted: 23 October 2019; Published: 4 November 2019

**Abstract:** In this paper, the influence of TiO$_2$ nanoparticle coating on cobalt-based electrodes was studied. Different coating treatment times were applied, and the results were compared to the hard-faced layer obtained with unmodified electrodes. The hard facing was done in three layers, the first being a Ni-based interlayer, followed by two layers of corrosion and wear-resistant Co-based Stellite 6 alloy. Pin-on-disc wear testing was applied, along with the metallographic study and hardness measurements of the hard-faced layers. Furthermore, energy-dispersive X-ray spectroscopy (EDS) analysis was conducted. It was found that the microstructural properties, as well as microhardness profiles, are modified in hard-faced layers obtained with modified electrodes. Interdendritic distances are altered, as are the dendrite growth directions. Titanium oxides are formed, which, along with the present complex carbides, increase the wear resistance of the hard-faced layers compared to layers obtained with untreated electrodes.

**Keywords:** hard facing; cobalt alloys; wear; nano-particle coating


## 1. Introduction

Cobalt-based alloys, commonly known as Stellites, are well-known wear and corrosion resisting alloys [1]. Stellites are Co-Cr-W-C alloys, containing around 30% chromium and 4% to 14% tungsten, while some alloys have their tungsten replaced with molybdenum for increased resistance to reductive media. Also, they contain up to around 2.5% of carbon, to form wear-resisting complex carbides [2–5]. Stellite alloys properties are the result of the combined effects of the relatively ductile Co-based matrix that supports relatively hard carbides, providing wear resistance at both room and elevated temperatures up to 600 °C [6,7]. The matrix is based on chromium and tungsten solid solution in cobalt, while the carbides are of complex nature, predominantly of the chromium-rich M$_7$C$_3$ eutectic type [8]. There is a variety of Stellite alloys, suited to various environments and applications [3]. Major Stellite alloys, their chemical composition, hardness, and applications are presented in Table 1.

**Table 1.** Major Stellite alloys with their chemical composition, hardness, and applications [9,10].

| Alloy | Co | Cr | W | C | Ni | Mo | Fe | Si | HRC | Properties | Applications |
|---|---|---|---|---|---|---|---|---|---|---|---|
| Stellite 1 | Base | 31 | 12 | 2.45 | <3.0 | <1.0 | <3.0 | <2.0 | 51–56 | Erosion, abrasion resistance, crack sensitive | Valve seat inserts, bearings, cutter edges |
| Stellite 6 | Base | 29 | 4 | 1.2 | <3.0 | <1.0 | <3.0 | <2.0 | 39–43 | General purpose wear, impact, cavitation resistance | Valve seats and gates, pump shafts and bearings, erosion shields, rolling couples |
| Stellite 12 | Base | 30 | 8 | 1.55 | <3.0 | <1.0 | <3.0 | <2.0 | 45–50 | High temperature and corrosion resistance, properties between Stellite 1 and 6 | Control plates, pump vanes, bearing bushes, narrow neck glass mold plungers, hard facing of engine valves, pinch rollers and rotor blade edges |
| Stellite 20 | Base | 32 | 16 | 2.45 | <3.0 | <1.0 | <3.0 | <2.0 | 53–57 | High abrasion, good corrosion resistance and low shock resistance. | Pump sleeves, rotary seal rings, wear pads, bearing sleeves, centerless grinder work rests |
| Stellite 21 | Base | 28 | - | 0.25 | 3 | 5.5 | <3.0 | <1.5 | 27–40 | Thermal, mechanical shock, cavitation resistance, not for abrasion | Building of forging and hot stamping dies |
| Stellite 25 | Base | 20 | 14 | 0.1 | 10 | <1.0 | <3.0 | <1.0 | 20–45 | Metal-on-metal wear, thermal fatigue, hot corrosion | Piercing points, forming tools, extrusion dies, furnace hardware |

The most common alloy is Stellite 6, owing this to its relatively balanced properties. It is sometimes regarded to be the standard Stellite alloy, with good wear resistance, corrosion, cavitation erosion, ductility and heat resistance, retaining these properties up to approximately 500 °C. The main corrosion mechanism is pitting, and it is well suited to be used in chloride solutions and seawater, where weight loss is usually under 0.05 mm per year at 22 °C [11]. More recently, Marques et al. [12] studied the application of Stellite alloys for fabricating of the components used in the production of second-generation ethanol. In such devices, biomass containing around 8% abrasive particles is loaded into the reactor, along with sugar cane biomass. To achieve improved abrasive wear properties, an increase in molybdenium content was proposed [13], due to the formation of reinforcing $Co_3Mo$ intermetallic compounds. However, molybdenium is a critical raw material (CRM) for the European Union. In addition to this, cobalt, tungsten, and silicon metal are all CRMs. Finally, Stellites contain chromium, that is near-CRM and nickel that is not CRM, but it is a relatively expensive metal [13,14]. That means, there is a considerable interest in increasing the life of Stellite components, to spare CRMs as much as possible. An alternative to adding molybdenium, and a possible way of increasing the life of Stellite components, is to enhance their wear properties. One of the potentially attractive ways of increasing the properties of the hard-faced layers made of Stellite 6 alloy is the introduction of $TiO_2$ nanoparticles into the existing electrode as proposed in [15–17]. In [15–17], where cellulose and rutile coated electrodes were modified by infiltrated $TiO_2$, this influenced the increased mechanical properties of the joints. In these studies, it was demonstrated that the addition of $TiO_2$ resulted in the increase in the number of complex Ti-Mn-Si oxide inclusions, acting as inoculants. That influenced the fine-grained acicular ferrite instead of Widmastaetten ferrite, which caused the increase in mechanical properties [15–17].

The aim of this study is to explore the influence of nanoparticles introduced on the Co-based hard-facing electrode with different immersion times on the wear resistance of the resulting hard-faced layer. The hard-faced layer was produced using the common shielded metal arc welding (SMAW) technique.

## 2. Materials and Methods

The SMAW (Shielded metal arc welding) process was used for hard facing. The base metal was S235JR, having the chemical composition shown in Table 2. Two types of hard-facing electrodes were used. The interlayer or buttering layer was done by Boehler Fox Nibas 70/20 (Boehler, Hamm, Germany; AWS A5.11-05: E NiCrFe-3) electrode in one pass. The final hard-faced layer was done by Co-based FSH Selectarc Co6, (Forges de Saint-Hippolyte, Roche-lez-Beaupre, France; Stellite 6; AWS A5.13: E CoCr-A) in two layers. Chemical compositions of the interlayer and hard-facing layer are presented in Tables 3 and 4. The overall thickness of the hard-faced layer was approximately 5 mm. The hard-facing parameters were in accordance with the electrode manufacturer's instructions (Table 5) and welding was done on an Iskra E10 (Iskra, Ljubljana, Slovenia) SMAW device.

**Table 2.** Chemical composition of S235JR (wt. %).

| C | Si | Mn | S | Cr | P | Al | Cu | Mo | Ni | Fe |
|---|----|----|----|----|----|----|----|----|----|----|
| 0.09 | 0.14 | 0.02 | 0.041 | 0.001 | 0.011 | 0.006 | 0.39 | 0.011 | 0.07 | balance |

**Table 3.** Nominal chemical composition of interlayer electrode AWS A5.11-05: E NiCrFe-3 (wt. %).

| C | Si | Mn | Cr | Fe | Mo | Nb | Co | Ni |
|---|----|----|----|----|----|----|----|----|
| 0.025 | 0.4 | 5.0 | 19.0 | 3.0 | 1.2 | 2.2 | 0.08 | balance |

**Table 4.** Nominal chemical composition of Co-based electrode AWS A5.13: E CoCr-A (wt. %)

| C | Si | Cr | W | Fe | Co |
|---|----|----|----|----|----|
| 1.1 | 1.0 | 28.0 | 4.5 | 3.0 | balance |

**Table 5.** Welding parameters.

| Electrode Core Diameter × Length | Ø 3.2 × 450 mm |
|---|---|
| Current type | DC+ |
| Welding current | 85 A |
| Welding voltage | 22 V |
| Welding speed | 12 cm/min |
| Electrode inclination along welding direction | 60° |

$TiO_2$ nanoparticles were introduced in the form of a coating on the hard-facing Co-based electrode. The distilled water solution of 20 nm hydrophilic $TiO_2$ nanoparticles (5 wt. %) was placed into the EMAG Emmi-5 (Emmi Ultrasonic, Moerfelden-Walldorf, Germany) ultrasonic bath, along with the fully submerged electrodes. Three immersion times were used: 1, 5, and 10 min, resulting in hard-faced specimens designated as 2, 3, and 4. These hard-faced layers were compared to the layer obtained with the untreated electrode (hard-faced specimen 1). After such treatment, the electrodes were dried for 1 h at 250 °C. Overall, four hard-faced specimens were prepared, which were water jet cut into Ø 10 mm specimens. One specimen was square, 30 mm wide, and used for metallographic examination and hardness measurement on its cross-section.

The pin-on-disc wear test was done on a customized Struers DP-U2 (Struers, Bellerup, Denmark) laboratory-polishing machine, with the polishing wheel replaced with SiC grinding paper. Three FEPA (Federation of European Producers of Abrasives) grit sizes were used: P240 (44.5–110 μm SiC grains, median grain size 58.5 μm, according to ISO 6344-1), P360 (29.6–87 μm, median grain size 40.5 μm) and P500 (21.5–77 μm, median grain size 30.2 μm). Three loadings were used: 700 g, 1000 g, and 1300 g. The specimens were placed into the brass holder and mounted in a Struers PdM-Force (Struers, Bellerup, Denmark) specimen mover. Prior to each wear test, the specimens were ground with P2000 SiC abrasive paper. The wheel spindle speed was 250 min$^{-1}$, wear time was 60 s, and the specimen axis was 70 mm away from the center of the spindle. During the wear test, a constant water flow of 100 mL/min was maintained, with a water temperature of 15 °C. The wear mass loss was reported for each test, with the mass of each specimen before and after wear measured by Tehtnica Type 2615 (Tehtnica, Zelezniki, Slovenia) analytic weight, having the accuracy of 0.1 mg. The results reported were calculated as an average of three specimens.

The metallographic characterization was conducted with a Leitz-Orthoplan (Leica-Leitz, Wetzlar, Germany) light microscope. Prior to this, the specimens were mounted, ground with abrasive papers (P150, P240, P360, P500, P600, P800, P1000, P1500, and P2000), and polished with diamond suspensions (6, 1 and $\frac{1}{4}$ μm diamond particles). The etching of the hard-faced layer was done by Aqua Regia (17% $HNO_3$, 50% HCl in glycerol), while the base metal was etched with Nital (3% $HNO_3$ in ethanol). Image analysis was done by ImageJ software. Furthermore, a JEOL JSM6460LV (JEOL, Tokyo, Japan) scanning electron microscope (SEM) equipped with Oxford Instruments INCA Microanalysis system EDS was used to evaluate the composition of different phases. Previously, metallographic specimens were coated with gold by a Ball-Tech Leica SCD-005 (Leica – Leitz, Wetzlar, Germany) device.

The HV5 Vickers hardness measurement was done according to ISO 6507-1:2005 on a VPM HPO 250 (VPM, Rauenstein, Germany) device, with a dwell time of 15 s. A hardness test was conducted in line from the specimen surface to the steel base material, with the distance between indentations of 0.5 mm. Five parallel linear measurements were done on each specimen, with the average value reported.

## 3. Results

### 3.1. Hardness Profiles

The results showing hardness profiles of hard-faced layers are shown in Figure 1. In all hard-faced specimens, hardness trends are similar, with values decreasing as the depth of the measurement is higher towards the nickel interlayer and finally the base metal. At a 5 mm distance from the surface, all hardness values reach a similar value which is the hardness of the base metal. There are several differences between the hardness profiles of tested specimens. The maximum hardness near the surface of the specimen 4 is the highest of all tested. Furthermore, the drop of hardness from the surface to the base metal is more pronounced in the specimen 1, hard-faced with unmodified electrodes.

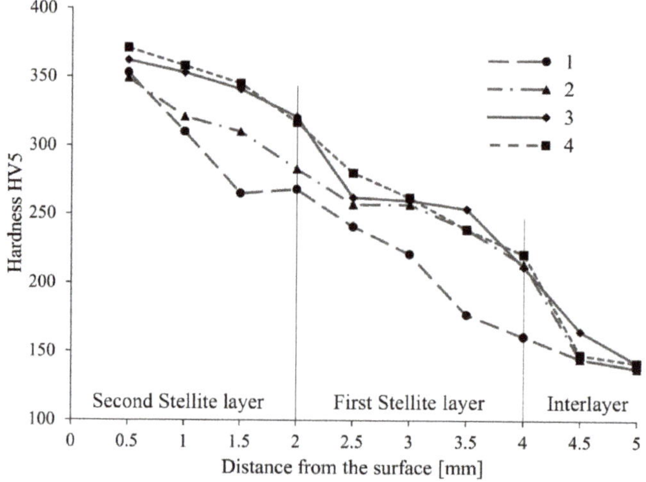


**Figure 1.** Hardness profiles of layers obtained with unmodified and modified electrodes.

### 3.2. Wear

The wear rate in the form of mass loss is shown in Figure 2. The coarser abrasive paper and a higher loading result in an increased mass loss of the hard-faced layer, which was expected. The modification of the electrodes by $TiO_2$ generally results in increased wear resistance of the resulting hard-faced layer, compared to the hard-faced layer obtained with unmodified electrodes. The general trend is that the mass loss is lower, that is, wear resistance is higher as the immersion time is longer. In other words, the advantage of the electrodes treated at longer times is higher as the abrasive grains are larger and as the loading is higher. This indicates that the nanoparticle coating on the electrode is beneficial for increased wear resistance of the hard-faced Stellite alloy.

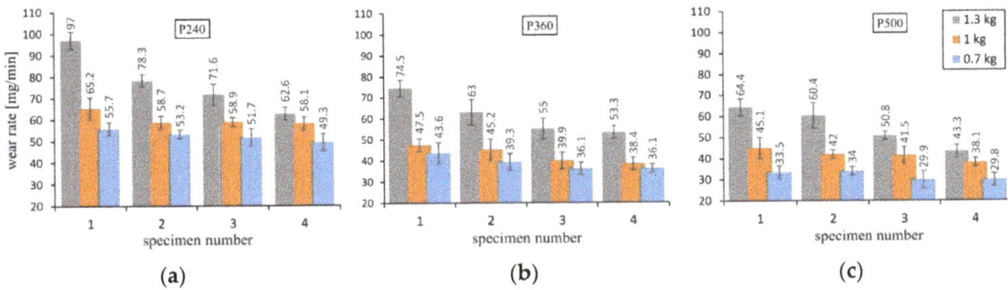


**Figure 2.** Pin-on-disc wear test results with the following SiC grit sizes: (**a**) P240; (**b**) P360; (**c**) P500

### 3.3. Microstructure

The microstructures obtained with the light microscope are given in Figures 3 and 4. Figure 3 depicts the microstructures of the hard-faced layer of different specimens, obtained with non-modified (1) and modified (2, 3, 4) at the surface, 1.5 mm under the surface, and 2.5 mm under the surface of the specimen. In all specimens, a typical microstructure consisting of the cobalt-based matrix and eutectic carbides typical for Stellite alloys is present.

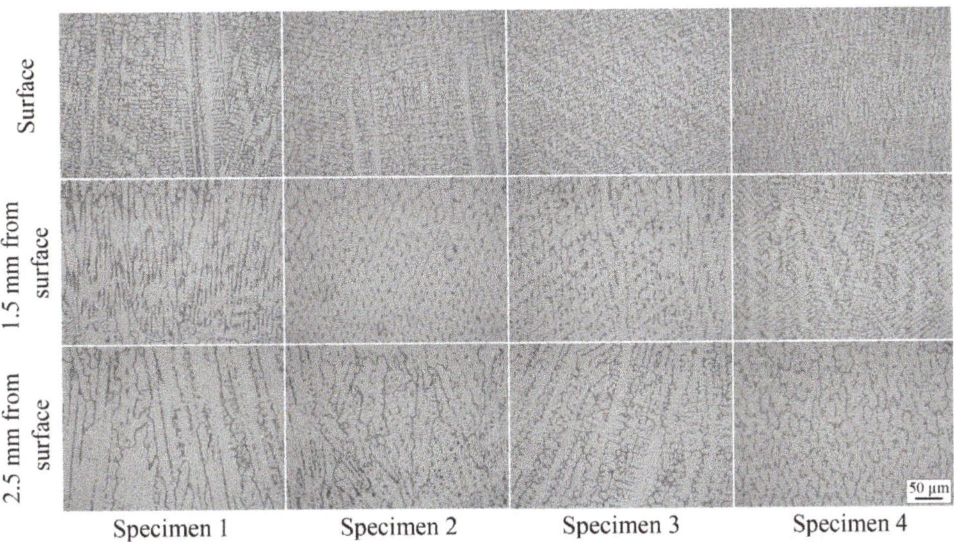


**Figure 3.** Microstructures of the hard-faced layer of different specimens at different depths.

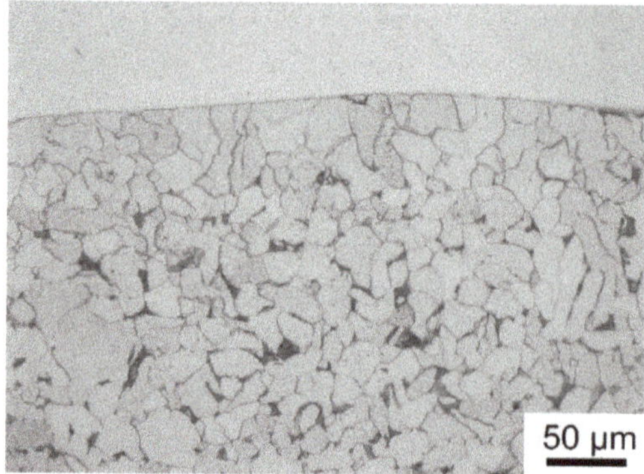

**Figure 4.** Microstructure of the S235JR base material near the fusion line.

In the surface layer, the interdendritic distance is higher in the hard-faced specimen obtained with untreated consumables (specimen 1) than those obtained with treated ones (specimens 2, 3, 4). Also, the carbide network in specimen 1 is opened to a higher extent than in other specimens, especially the specimen 4, where it is almost completely closed.

A typical cast dendritic primary microstructure morphology gradually becomes less pronounced in terms of the presence of interdendritic phases as the depth is increased. Interdendritic phases morphology becomes gradually changed from a typical interdendritic form to an intergranular and isolated form as the depth increases. However, this transition does not appear to be equal in all specimens. The most pronounced transition is in the specimen 1, obtained with unmodified electrodes.

The results shown in Table 6 reveal that the interdendritic phase (dark areas) volume fraction becomes lower as the depth is increased, which is in accordance with microstructures given in Figure 3. Also, the volume fraction of the interdendritic phases in the specimen hard-faced with treated electrodes is notably smaller than those in specimens hard-faced with treated electrodes, the highest being in specimen 4.

**Table 6.** Interdendritic phase (dark areas) volume fraction in microstructure.

| Specimen | Interdendritic Phase Volume Fraction [%] | | |
|:---:|:---:|:---:|:---:|
| | **Surface** | **1.5 mm under the Surface** | **2.5 mm under the Surface** |
| 1 | 12.0 | 11.5 | 7.9 |
| 2 | 12.3 | 12.0 | 10.3 |
| 3 | 12.5 | 12.1 | 10.1 |
| 4 | 14.0 | 12.4 | 11.0 |

The microstructure of the base metal is typical for S235JR steel, with slight recrystallization at the fusion line, as shown in Figure 4. Such behavior is noted for all tested specimens.

*3.4. EDS Analysis*

The results of the EDS analysis are shown in Figures 5 and 6. In Figures 5 and 6, the EDS analyses were done in the area near the surface of the hard-faced layer of specimens 1 and 4, respectively. The interdendritic carbides in specimen 1 belong to the $M_7C_3$ type of the complex nature (Cr–W) with predominant chromium content, while the matrix is Co-based, as shown in Figure 5. However, in specimen 4, there are also islands that besides the elements detected in specimen 1, contain titanium and oxygen, most probably of titanium-oxide nature. The matrix in specimen 4 has a similar chemical composition as the one in specimen 1, indicating that the matrix does not undergo modification related to chemical content in the presence of nanoparticles.

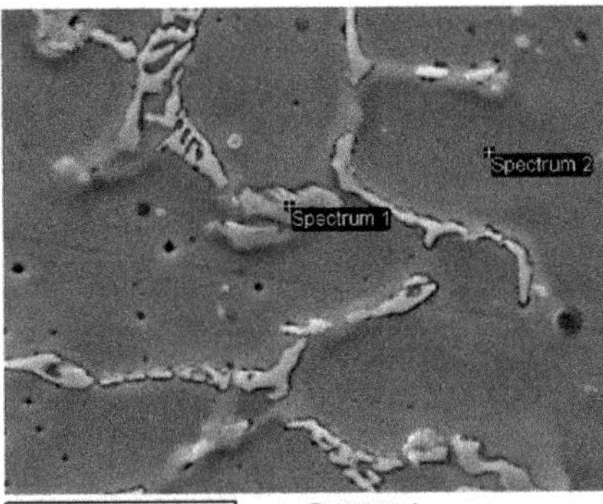

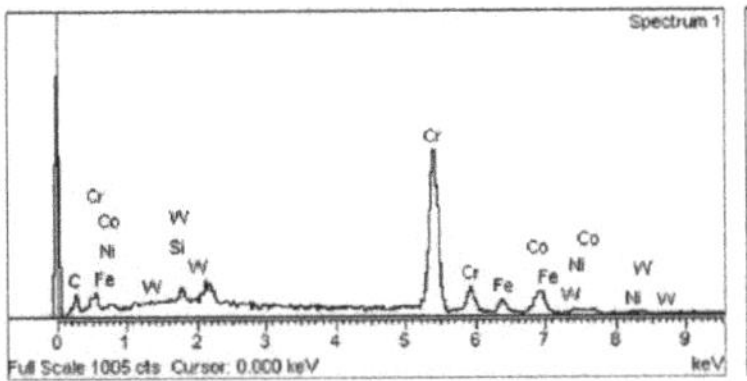

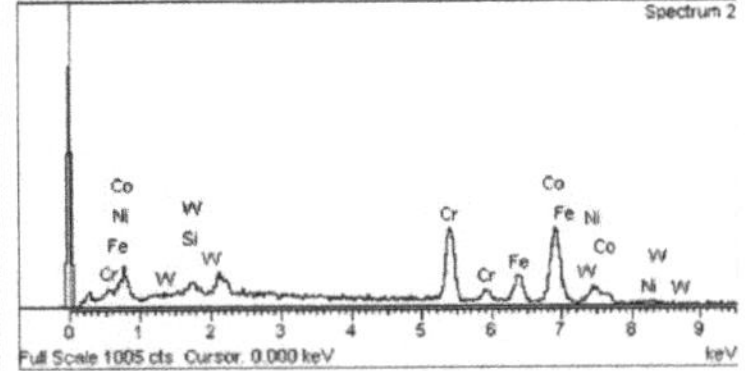

| | Atomic % | |
|---|---|---|
| | **Spectrum 1** | **Spectrum 2** |
| **C** | 58.46 | 0 |
| **Si** | 0 | 1.98 |
| **Cr** | 29.42 | 27.54 |
| **Fe** | 3.15 | 14.52 |
| **Co** | 7.22 | 44.50 |
| **Ni** | 1.15 | 10.85 |
| **W** | 0.8 | 0.61 |

**Figure 5.** EDS analyses of carbides and base metal in the untreated specimen 1.

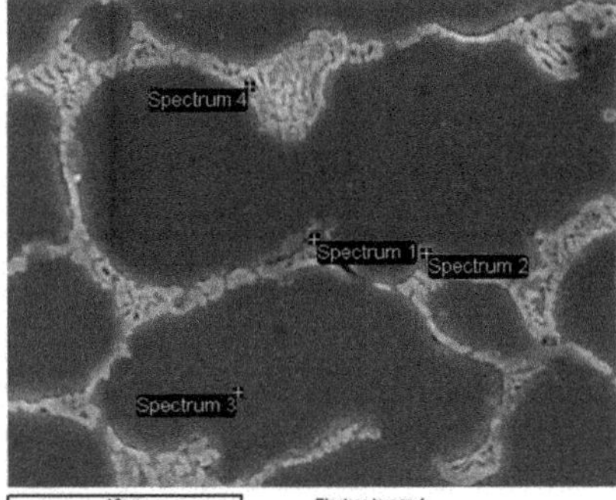

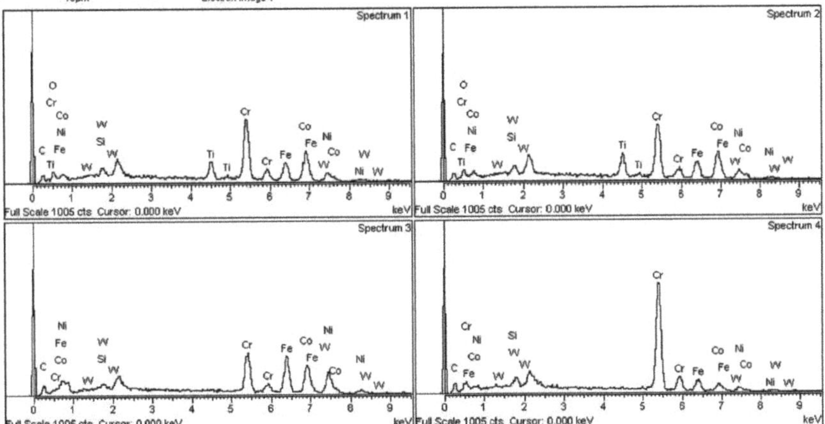

| | Atomic % | | | |
|---|---|---|---|---|
| | **Spectrum 1** | **Spectrum 2** | **Spectrum 3** | **Spectrum 4** |
| **C** | 39.98 | 40.45 | 2.54 | 53.69 |
| **Si** | 1.24 | 0.67 | 2.80 | 1.21 |
| **Cr** | 16.57 | 16.05 | 16.05 | 30.11 |
| **Fe** | 6.83 | 6.79 | 18.21 | 2.98 |
| **Co** | 12.42 | 12.66 | 45.18 | 6.81 |
| **Ni** | 4.35 | 4.23 | 14.62 | 4.50 |
| **W** | 0.48 | 0.81 | 0.6 | 0.7 |
| **O** | 14.24 | 13.42 | 0 | 0 |
| **Ti** | 3.9 | 4.91 | 0 | 0 |

**Figure 6.** EDS analyses of oxides, carbides and base metal in specimen 4 obtained with electrode treated with $TiO_2$ nanoparticles (10 mm immersion).

## 4. Discussion

In this paper, the immersion technique was used to create an additional coating consisting of $TiO_2$ for increasing the performance of SMAW electrodes for cobalt alloy hard facing, more specifically Stellite 6.

The addition of nanoparticle-based coating to the existing SMAW electrode results in an increased abrasive wear resistance for all hard-faced specimens obtained with modified electrodes. This can be attributed to several effects. The first effect is the occurrence of oxides, the second the decreased interdendritic distance, and the third is the increase in hardness in the surface and subsurface layers. The increase in hardness can also be correlated to the occurrence of titanium-oxides [18–20]. These oxides occur in the form of blocky structures, usually between the Co-based matrix and the complex $M_7C_3$ chromium-based carbides. Titanium-oxides add to the wear resistance of the resulting hard-faced layer. During welding, a relatively high temperature is developed, sufficient to cause the dissociation of $TiO_2$ in accordance with the following chemical reaction:

$$TiO_2 \rightarrow Ti + O_2 \tag{1}$$

After cooling, the oxygen again tends to form oxides, primarily with Ti. As shown in [15], their solidification temperatures are relatively high: $Ti_2O_3$ 2130 °C; $TiO_2$ 1843 °C [21]. A common effect was noticed in cellulose and rutile electrode weldments where similarly modified electrodes influenced the increased inoculation of acicular ferrite through the mechanism of an increased inclusion formation [15,22–26]. However, in this case, the solidification continues with dendritic inoculation and growth.

During cooling, dendrites (Co-alloy forming the metal matrix) solidify second, after titanium oxides at 1475 °C. Finally, the complex (predominantly chromium) carbides of $M_7C_3$ type are left to solidify the last at 1175 °C [27], forming the microstructure that defines the hard-faced layer obtained with the modified electrodes. This can explain the more random orientation of dendrites shown in Figures 3d and 4d (specimen 4) compared to the specimen 1 obtained with the unmodified electrode. Thus, titanium-oxides have multiple effects, beginning with increasing the hardness both at the specimen surface and in depth, as well as decreasing the interdendritic distance, making the carbide network finer, causing a more effective dispersion strengthening of the matrix, as explained in [28,29].

The increase in wear resistance by adding a $TiO_2$ nanoparticle coating to the electrodes can influence the increase in wear resistance of the hard-faced layer. This is beneficial from the point of view of CRMs since the hard-faced layer contains cobalt as a basis, tungsten, silicon metal, and molybdenum, as well as near-CRM such as chromium and expensive materials such as nickel. Increased wear resistance can influence the reduction in consumption of these elements, lowering the dependence on them.

## 5. Conclusions

Based on the results shown in this work, the following conclusions can be drawn:

- The hardness values of hard-faced layers obtained with $TiO_2$ nanoparticle coating on the SMAW electrode are higher than those of the layers obtained with untreated electrodes.
- The drop of hardness towards the depth of the hard-faced layer is less pronounced in layers obtained with nanoparticle coated electrodes.
- The wear resistance of layers obtained with nanoparticle coatings is increased compared to specimens obtained with unmodified electrodes.
- The highest wear resistance was obtained with electrodes immersed for the longest period of time, 10 min. This effect can be observed for all tested abrasive grain sizes (P240–P500) and loadings (0.7–1.3 kg). The most pronounced increase in wear resistance was when coarser grains at higher loadings were applied.

- Several effects influence the increase in wear resistance: the formation of blocky titanium-oxides, the rise in carbide and oxide volume fraction, and the refinement of dendrites. The dendrite network becomes almost completely closed which can also be one of the effects beneficial for the increase in wear.
- The increased wear resistance of specimens obtained with modified electrodes influence the prolonged life of the component, which can provide significant savings in CRMs such as cobalt, tungsten, molybdenium, and silicon metal, present in the hard-faced alloy.

**Author Contributions:** Conceptualization, methodology and writing S.B.; investigation P.J. and M.D.; investigation D.L.Z.; formal analysis B.P.; review and editing P.H. and L.J.

**Funding:** This research received no external funding.

**Acknowledgments:** This article is based on work from COST Action "Solutions for Critical Raw Materials under Extreme Conditions", supported by COST (European Cooperation in Science and Technology).

**Conflicts of Interest:** The authors declare no conflict of interest.

## References


1. Haynes, E. Metal Alloy. U.S. Patent No. 873,745, 17 December 1907.
2. Drapier, J.M.; Davin, A.; Magnee, A.; Coutsouradis, D.; Habraken, L. Abrasion and corrosion resistant cobalt base alloys for hardfacing. *Wear* **1975**, *33*, 271–282. [CrossRef]
3. Wood, P.D.; Evans, H.E.; Ponton, C.B. Investigation into the wear behaviour of Stellite 6 during rotation as an unlubricated bearing at 600 °C. *Tribol. Int.* **2011**, *44*, 1589–1597. [CrossRef]
4. Gholipour, A.; Shamanian, M.; Ashrafizadeh, F. Microstructure and wear behaviour of Stellite 6 cladding on 174 PH stainless steel. *J. Alloys Compd.* **2011**, *509*, 4905–4909. [CrossRef]
5. Da Silva, W.S.; Souza, R.M.; Mello, J.D.B.; Goldenstein, H. Room temperature mechanical properties and tribology of NICRALC and Stellite casting alloys. *Wear* **2011**, *271*, 1819–1827. [CrossRef]
6. Berns, H. Microstructural properties of wear-resistant alloys. *Wear* **1995**, *181*, 271–279. [CrossRef]
7. Yao, M.X.; Wu, J.B.C.; Xu, W.; Liu, R. Metallographic study and wear resistance of a high-C wrought Co-based alloy Stellite 706K. *Mater. Sci. Eng. A Struct.* **2005**, *407*, 291–298. [CrossRef]
8. Davis, J.R. *Nickel, Cobalt, Their Alloys*; ASM International: Cleveland, OH, USA, 2001.
9. Exocor Data Sheet. Available online: http://exocor.com (accessed on 10 August 2019).
10. Kennametal Data Sheet. Available online: https://kennametal.com (accessed on 10 August 2019).
11. Deloro Stellite 6 Data Sheet. Available online: http://www.deloro.com (accessed on 14 February 2017).
12. Yao, M.X.; Wu, J.B.C.; Xie, Y. Wear, corrosion and cracking resistance of some W- or Mo-containing Stellite hardfacing alloys. *Mater. Sci. Eng. A Struct.* **2005**, *407*, 234–244. [CrossRef]
13. Marques, F.P.; Bozzi, A.C.; Scandian, C.; Tschiptschin, A.P. Microabrasion of three experimental cobalt-chromium alloys: Wear rates and wear mechanisms. *Wear* **2017**, *390*, 176–183. [CrossRef]
14. Luisa Grilli, M.; Bellezze, T.; Gamsjäger, E.; Rinaldi, A.; Novak, P.; Balos, S.; Piticescu, R.R.; Letizia Ruello, M. Solutions for critical raw materials under extreme conditions: A review. *Materials* **2017**, *10*, 285. [CrossRef] [PubMed]
15. Report on Critical Raw Materials for EU, Report of the Ad-Hoc Working Group on Defining Critical Raw Materials for EU. May 2014. Available online: http://mima.geus.dk/report-on-critical-raw-materialsen.pdf (accessed on 28 August 2019).
16. Balos, S.; Sidjanin, L.; Dramicanin, M.; Labus Zlatanovic, D.; Pilic, B.; Jovicic, M. Modification of cellulose and rutile welding electrode coating by infiltrated $TiO_2$ nanoparticles. *Met. Mater. Int.* **2016**, *22*, 509–518. [CrossRef]
17. Balos, S.; Dramicanin, M.; Labus Zlatanovic, D.; Sidjanin, L. SMAW welding aided by $TiO_2$ nano particles. In Proceedings of the 3rd IIW Sout-East European Welding Congress, Timisoara, Romania, 3–5 June 2015; pp. 1–6.
18. Balos, S.; Dramicanin, M.; Labus Zlatanovic, S.; Sidjanin, L.; Pilic, B. Rutile Electrodes Enhanced with $TiO_2$ nanoparticles. In *Advanced Materials Research*; Trans Tech Publications: Timisoara, Romania, 2016; pp. 69–74.

19. Fattahi, M.; Nabhani, N.; Vaezi, M.R.; Rahimi, E. Improvement of impact toughness of AWS E6010 weld metal by adding $TiO_2$ nanoparticles to the electrode coating. *Mater. Sci. Eng. A Struct.* **2011**, *528*, 8031–8039. [CrossRef]

20. Byun, J.-S.; Shim, J.-H.; Suh, J.-Y.; Oh, Y.-J.; Cho, Y.W.; Shim, J.-D.; Lee, D.N. Inoculated acicular ferrite microstructure and mechanical properties. *Mater. Sci. Eng. A Struct.* **2011**, *319*, 326–331.

21. Greenwood, N.; Earnshaw, A. *Chemistry of the Elements*, 2nd ed.; Butterworth-Heinemann: Woburn, UK, 1997; pp. 961–962.

22. Hossein Nedjad, S.; Farzaneh, A. Formation of fine intragranular ferrite in cast plain carbon steel inoculated by titanium oxide nanopowder. *Scr. Mater.* **2007**, *57*, 937–940. [CrossRef]

23. Kiviö, M.; Holappa, L.; Iung, T. Addition of dispersoid titanium oxide inclusions in steel and their influence on grain refinement. *Metall. Mater. Trans. B* **2010**, *41*, 1194–1204. [CrossRef]

24. St-Laurent, S.; L'Esperance, G. Effects of chemistry, density and size distribution of inclusions on the nucleation of acicular ferrite of C-Mn steel shielded-metal-arc-welding weldments. *Mater. Sci. Eng. A Struct.* **1992**, *149*, 203–216. [CrossRef]

25. Zhang, D.; Terasaki, H.; Komizo, Y. In situ observation of the formation of intragranular acicular ferrite at non-metallic inclusions in C–Mn steel. *Acta Mater.* **2010**, *58*, 1369–1378. [CrossRef]

26. Byun, J.S.; Shim, J.H.; Cho, Y.W. Influence of Mn on microstructural evolution in Ti-killed C–Mn steel. *Scr. Mater.* **2003**, *48*, 449–454. [CrossRef]

27. Davis, J.R. *Alloying: Understanding the Basics*; ASM International: Cleveland, OH, USA, 2001.

28. Smallman, R.; Bishop, R.J. *Modern Physical Metallurgy and Materials Engineering*, 6th ed.; Butterworth Heinemann: Woburn, UK, 1999.

29. Sun, N.; Apelian, D. Friction stir processing of aluminum cast alloys for high performance applications. *JOM* **2011**, *63*, 44–50. [CrossRef]

*Article*

# CaO–CaZrO₃ Mixed Oxides Prepared by Auto–Combustion for High Temperature CO$_2$ Capture: The Effect of CaO Content on Cycle Stability

**Igor Luisetto *, Maria Rita Mancini, Livia Della Seta, Rosa Chierchia, Giuseppina Vanga, Maria Luisa Grilli and Stefano Stendardo ***

Italian National Agency for New Technologies, Energy and Sustainable Economic Development (ENEA), Via Anguillarese 301, 00123 Rome, Italy; rita.mancini@enea.it (M.R.M.); livia.dellaseta@enea.it (L.D.S.); rosa.chierchia@enea.it (R.C.); giuseppina.vanga@enea.it (G.V.); marialuisa.grilli@enea.it (M.L.G.)

* Correspondence: igor.luisetto@enea.it (I.L.); stefano.stendardo@enea.it (S.S.); Tel.: +39-3048-3255 (I.L.); +39-3048-4494 (S.S.)

Received: 5 April 2020; Accepted: 3 June 2020; Published: 5 June 2020

**Abstract:** Cycling high temperature CO$_2$ capture using CaO–based solid sorbents, known as the calcium looping (CaL) process, is gaining considerable scientific and industrial interest due to the high theoretical sorbent capacity (0.78 gCO$_2$/gCaO), the low specific cost, and the negligible environmental impact of the employed materials. In this work, we investigated the self–combustion synthesis of CaO–CaZrO₃ sorbents with different CaO contents (40, 60, and 80 wt%) for use in the CaL process. CaZrO₃ was used as a spacer to avoid CaO grains sintering at high temperature and to reduce the diffusional resistances of CO$_2$ migrating towards the inner grains of the synthetic sorbent. Samples were characterized by X–ray diffraction (XRD), Brunauer–Emmett–Teller (BET), and scanning electron microscopy (SEM) analyses. The reaction between CO$_2$ and CaO (i.e., carbonation) was carried out in 20 vol% CO$_2$ at 650 °C and calcination (i.e., decomposition of CaCO₃ to CaO and CO$_2$) at 900 °C in pure Ar or with 85 vol% CO$_2$ using a thermogravimetric analyzer (thermogravimetric/differential thermal analysis (TG–DTA)). The most stable sorbent was with 40 wt% of CaO showing a CO$_2$ uptake of up to 0.31 g CO$_2$/g$_{sorbent}$ and 0.26 g CO$_2$/g$_{sorbent}$ operating under mild and severe conditions, respectively. The experimental data corroborated the prediction of the shrinking core spherical model in the first phase of the carbonation. A maximum reaction rate of 0.12–0.13 min$^{-1}$ was evaluated in the first cycle under mild and severe conditions of regeneration.

**Keywords:** CO$_2$ capture; calcium looping; nanometric CaZrO₃ particles

## 1. Introduction

Among different sorbents, metal oxides contained in naturally occurring minerals represent the cheapest option for the CO$_2$ capture process; the calcined forms of limestone and dolomite [1–3] show a high reactivity within a temperature range from 550 to 750 °C, which largely fits with reforming, gasification, and pyrolysis processes, and may be regenerated by thermal decomposition of carbonates at temperatures ranging from 850 to 900 °C. This process can be implemented for CO$_2$ removal from product gases in fluidized bed systems for combustion gasification, or methane steam reforming with water–gas shift reaction [4–6]. In this process, CaO is converted into CaCO₃ during the CO$_2$ uptake (carbonation Equation (1)).

$$CaO + CO_2 \rightleftharpoons CaCO_3 \quad \Delta H^0 = -178 \text{ kJ·mol}^{-1} \tag{1}$$

The spent solid sorbent is subsequently regenerated by releasing $CO_2$ in a calcination step. It is desirable to produce a concentrated $CO_2$ stream during the regeneration process and, in this case, the calcination will often be carried out in an atmosphere rich in $CO_2$. This puts requirements on the sorbent, e.g., the sorbent must withstand a high $CO_2$ concentration during its regeneration for an extended number of cycles. However, the presence of $CO_2$ during the regeneration step has been shown to decrease the calcination rate of the solid sorbent, and also to cause sintering of CaO grains with associated negative effects on the $CO_2$ uptake capacity after multiple cycles [7–9]. In order to lower the molar fraction of $CO_2$ during calcination, it has been proposed to add steam that is easily separable after the calcination step [10]. Besides the cost for producing and separating the steam, steam itself was also found to cause sintering of CaO, thereby reducing the $CO_2$ uptake capacity of the regenerated sorbent; steam can lead to changes in the pore–size distribution (PSD) through the $CaCO_3$ layer [11]. In any case, the sorbent kinetics and stability of $CO_2$ uptake capacity during cycling in industrial–sized plants are key issues and strongly influence the cost of the CaL technology [12]. The sorbent must thus show a high stability and stable sorption capacity throughout multiple $CO_2$ uptake–regeneration cycles. Further important properties of a good sorbent are as follows: low–cost manufacturing, fast reaction kinetics, as well as mechanical stability and sintering resistance. Although natural carbonates, e.g., limestone ($CaCO_3$) or dolomite ($[Ca,Mg]CO_3$), seem to be the best candidates given their low cost and high availability, they show a pronounced decay in $CO_2$ sorption capacity during repeated cycling [1,9]. The carbonation of these materials is characterized by an abrupt transition from the rapid initial reaction rate, most likely kinetically controlled, to a slower reaction rate, which is most likely diffusion controlled [3,8,13]. Other drawbacks of these materials are as follows: sorption capacity decay, due to sintering and modification of the particles' porous structure [14–16], and low mechanical resistance under fluidization conditions, specifically for the calcined state, which may therefore cause elutriation of fine particles. To counteract this, efforts have been made to enhance the performances of naturally occurring sorbents. Thermal pretreatment of natural dolomite or limestone [17–19] and hydration with steam [20–22] improve the $CO_2$ sorption capacity to some extent during subsequent cycling. Interestingly, thermally pretreated limestone sometimes shows the effect of self–activation during cycling, i.e., the sorption capacity increases during cycling [9,17]. Although positive results were obtained, these tests have frequently used regeneration in a 100% nitrogen atmosphere without taking into account the effects from $CO_2$ on sintering. However, it is important to highlight the fact that, when dolomite and limestone are calcined in a $CO_2$–containing atmosphere, sintering processes occur [23–25]. These drawbacks counteract the advantage of the low cost of naturally occurring sorbents [26,27]. Alternatives to Ca–based materials are solid sorbents based on both alkali–promoted alumina, hydrotalcites, and some other Mg–based material [28,29]. The operation temperatures of these materials are in the range of 300–600 °C.

The sorbent materials initially chosen for proof–of–concept in the CaL process had high capacity, but low mechanical stability. To overcome the abovementioned limitations of the solid sorbents that hamper the exploitation of sorption–enhanced processes, the research community has focused its efforts on the development of synthetic materials (or, alternatively, properly modified natural minerals) able to keep a very high $CO_2$ uptake capacity over a higher number of sorption/regeneration cycles and compatible with the use in a fluidized bed reactor. The authors in [30] investigated the effects of thermal pretreatment on a naturally occurring sorbent to enhance the stability of the modified sorbent. In addition to the attempts to modify naturally occurring minerals to improve their properties, there is also the option to develop completely new synthetic sorbents with tailored properties. Several sorbents based on CaO have been synthesized, including high surface precipitated calcium carbonate [31] mixed oxides with the formula $Ca_{0.9}M_{0.1}O_x$, where M = Cu, Cr, Co, Mn [32]; CaO sorbents derived from organic salt precursors [33]; CaO doped with Cs [34]; sorbents derived from calcium lignosulphonate [35] or sorbents based on $Na_2CO_3$ or NaCl [14,36]; CaO–based sorbents using calcium aluminate ($Ca_{12}Al_{14}O_{33}$) as a stable matrix [9,37,38]; and CaO–based sorbents using calcium titanate ($CaTiO_3$)–based perovskite as binder [39], which showed good stable $CO_2$ uptake capacity during cycling tests.

The aim of this paper was to further investigate Ca–based sorbent materials using calcium zirconate ($CaZrO_3$) perovskite as binder. The new CaO–based material was tested under relevant conditions in several multi–cycle TG–DTA tests and characterized by SEM, XRD, and BET analyses.

## 2. Materials and Methods

All the chemical reagents were purchased from Merck Italy and used as received. Samples were prepared using the auto–combustion method, i.e., stoichiometric amounts of metal nitrate ($Ca(NO_3)·H_2O$ and $ZrO(NO_3)_2·5H_2O$) and glycine ($NH_2CH_2COOH$) as fuel with $\frac{NO_3^{2-}}{Glycine} = 0.9$ were dissolved in $H_2O$. The temperature of the solution was measured with a type K thermocouple (RS–PRO, Italy). At the beginning, the temperature was increased to 80 °C, favoring the $H_2O$ evaporation to trigger the auto–combustion. Then, the temperature was gradually increased with a ramp of 15 °C min$^{-1}$ up to 250 °C. After a short time, which increased from ~3 to ~10 min as the molar ratio between Ca and Zr decreased, the self–combustion occurred with a rapid flame propagation that ended after a few minutes, producing very fine and voluminous powders that were calcined in air at 900 °C for 5 h to remove carbonaceous impurities. The samples investigated in this work were named as follows: $CaZrO_3$ (CZ), 40 wt% $CaO–CaZrO_3$ (CCZ40), 60 wt% $CaO–CaZrO_3$ (CCZ60), and 80 wt% $CaO–CaZrO_3$ (CCZ80).

XRD measurements were carried out on a RIGAKU SmartLab system (RIGAKU supplier: Assing spa, Italy) equipped with a Cu K$\alpha$ ($\lambda = 0.15418$ nm) source for powder materials. The reference intensity ratio (RIR) method was used for quantitative phase analysis [40]. Because all phases were identified, no preferred texturization or amorphous phase was detected, and all RIR values were known for each phase, quantitative analysis was then permitted without adding any standard to the unknown specimen. The crystallite size was estimated from the $CaZrO_3$ (121), CaO (111), and $Ca(OH)_2$ (101) reflection using the Scherrer Equation (2), where the shape factor K is equal to 0.9, $\lambda$ *is* the wavelength of the X–ray radiation used, $\beta$ is the full width at half maximum intensity (FWHM), and $\theta$ is the Bragg angle of the corresponding reflection:

$$d(nm) = K\lambda/(\beta·\cos\theta) \qquad (2)$$

$N_2$ adsorption/desorption isotherms were measured at $-196$ °C using a Micromeritics ASAP 2020 adsorption apparatus (Micromeritics supplier: Alfatest srl, Italy). Samples were degassed at 350 °C under vacuum for 4 h before analysis. The specific surface area (SSA) was determined using the BET equation method in the linear range of relative pressure ($p/p^0$) of 0.07–0.3. The pore–size distributions were calculated from desorption branches using the Barrett–Joyner–Halenda (BJH) method. The total pore volume was estimated to be the liquid $N_2$ volume at relative pressure $p/p^0$ of 0.95.

Morphological characterization of sorbent materials was performed by scanning electron microscopy (SEM) using a Tescan Vega 3 with LaB6 filament (Tescan supplier: Assing spa, Italy).

Sorbent analysis were carried out in a Netzsch STA 449C Jupiter thermo–microbalance–based simultaneous thermogravimetric/differential scanning calorimetry (TG–DSC) system (with integrated mass flow controller TG resolution: 0.1 µg) (Netzsch, Germany). An amount of sorbent (ranging from 20 to 30 mg) was placed in an alumina crucible. Before the carbonation test, each sorbent was first pretreated up to 900 °C at a ramp of 10 °C·min$^{-1}$ under 100% Ar gas flow for 20 min in order to remove all impurities and pre–adsorbed $CO_2$. The $CO_2$ uptake capacity of the sorbents was tested under mild and severe conditions as follows:

(a) under mild conditions, the carbonation occurred at 650 °C for 20 min in 20 vol% $CO_2$ in Ar ($CO_2$ flow of 20 cm$^3$·min$^{-1}$ and Ar flow of 80 cm$^3$·min$^{-1}$) followed by the calcination step under 100% Ar gas flow for 20 min at 900 °C. All heating and cooling procedures of the experiment were performed in Ar flow of 100 cm$^3$/min with the heating or cooling rate of 10 °C·min$^{-1}$.

(b) under severe conditions, the carbonation occurred at 650 °C for 20 min in 20 vol% $CO_2$ in Ar ($CO_2$ flow of 20 cm$^3$·min$^{-1}$ and Ar flow of 80 cm$^3$·min$^{-1}$) followed by the calcination step under 80 vol% $CO_2$ ($CO_2$ flow of 80 cm$^3$·min$^{-1}$ and Ar flow of 20 cm$^3$·min$^{-1}$) for 20 min at 900 °C. After the calcination step, the cooling procedures of the experiment were performed in Ar flow of 100 cm$^3$/min with cooling rate of 10 °C·min$^{-1}$.

The performance and stabilities of the sorbents were evaluated through 12 carbonation–calcination cycles. The $CO_2$ uptake per gram of sorbent, $C_n$, and the CaO conversion (%), $X_n$, were calculated according to Equations (3) and (4):

$$C_n = \frac{m^n_{max} - m^n_{min}}{m^n_{min}} \tag{3}$$

$$X_n = \frac{C_n}{\eta} \times \frac{M_{CaO}}{M_{CO_2}} \times 100 \tag{4}$$

where $m^n_{max}$ is the mass of the sorbent at the end of cycle $n$ at 650 °C under mild and severe conditions or at 850 °C during the subsequent ramp temperature under severe conditions, $m^n_{min}$ is the initial mass of the sorbent at cycle $n$ at 650 °C under mild and severe conditions, $M_{CaO}$ is the molecular weight of CaO, $M_{CaCO_3}$ is the molecular weight of $CaCO_3$, and finally $\eta$ is the CaO mass fraction of the sorbent.

## 3. Results and Discussion

### 3.1. Sorbent Characterization

#### 3.1.1. X–Ray Diffraction Characterization of the As–Prepared Sorbents

The X–ray diffraction patterns of samples calcined at 900 °C are shown in Figure 1 and the main crystalline phases are reported in Table 1.

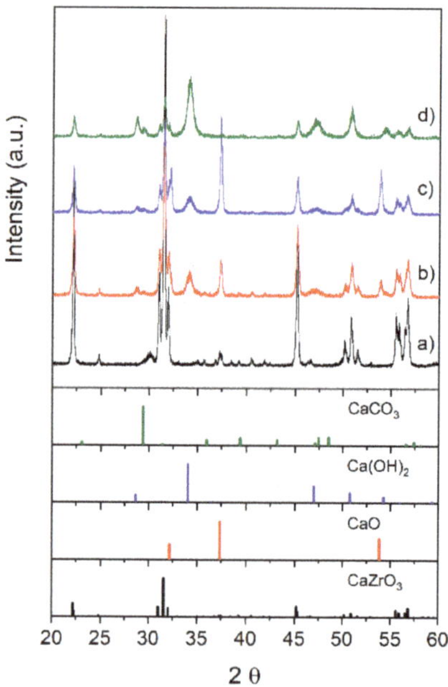


**Figure 1.** X–ray diffraction (XRD) patterns of as–prepared samples: (a) CZ, (b) CCZ40, (c) CCZ60, and (d) CCZ80. The position and intensity of the peaks corresponding to the main phases present are also included.

The CZ sample, containing a stoichiometric ratio between Ca and Zr, consisted of a perovskite phase with orthorhombic symmetry characteristic of $CaZrO_3$ oxide (JCPDS card no. 35–0790). However, the presence of impurity phases, such as $Ca_{0.15}Zr_{0.85}O_{1.85}$ (JCPDS card no. 26–0341), CaO (JCPDS card no. 37–1497), $Ca(OH)_2$ (JCPDS card no. 01–1079), and $CaCO_3$ (JCPDS card no. 47–1743), was also observed. As Ca content increased, the intensity of the $CaZrO_3$ perovskite phase decreased and the diffraction peaks became wider, suggesting the progressive decrease of the crystallite size. In addition to the perovskite phase, CCZ40 and CCZ60 samples showed mainly peaks attributed to the CaO and $Ca(OH)_2$ phases, whereas the CCZ80 sample showed principally peaks of $Ca(OH)_2$. The presence of the $Ca(OH)_2$ phase, which originated from the reaction of CaO with ambient humidity during material storage and handling, reflects the high surface reactivity of these mixed oxides. The high reactivity with the ambient air was also highlighted by the formation of $CaCO_3$, which was 9.5% of the phases in CCZ80. However, $Ca(OH)_2$ and $CaCO_3$ were completely converted to CaO during the samples' preheating step at 900 °C (see the TG results in Figure 2). As reported in Table 1, the crystalline size, estimated by the Scherrer equation, of $CaZrO_3$ perovskite decreased progressively from 63.9 nm to about 32 nm from CZ to CCZ80 because the increased amount of calcium species reduced the sintering of $CaZrO_3$ during sample calcination. The crystalline size of $Ca(OH)_2$ progressively decreased from 16.0 nm in CCZ40 to 11.2 nm in CCZ60 samples; similarly, CaO decreased from 35.5 nm to 31.2 nm for CCZ40 and CCZ60 samples, respectively. Therefore, the progressive increase of the Ca/Zr ratio during the synthesis generates crystallites of small dimensions.

**Table 1.** Crystallite size, phase amount, and textural properties of as–prepared samples.

| Sample | Crystallite Size (nm) [1] | | | Phase Amount (%) | | | | SSA $(m^2 \cdot g^{-1})$ | Pore Volume $(cm^3 \cdot g^{-1})$ |
|---|---|---|---|---|---|---|---|---|---|
| | $CaZrO_3$ | CaO | $Ca(OH)_2$ | $CaZrO_3$ | CaO | $Ca(OH)_2$ | $CaCO_3$ | | |
| CZ | 63.9 | – | – | 89.1[2] | 6.9 | 3 | 1.0 | 11.2 | 0.043 |
| CCZ40 | 56.4 | 35.5 | 16.0 | 52.8 | 20.4 | 25.5 | 0.3 | 32.4 | 0.095 |
| CCZ60 | 30.5 | 31.2 | 12.2 | 35.8 | 44.8 | 18.9 | 0.5 | 52.7 | 0.160 |
| CCZ80 | 32.3 | – | 11.2 | 14.0 | 3.0 | 73.5 | 9.5 | 48.0 | 0.123 |

[1] Calculated according to the Scherrer equation. [2] The phase amount of $Ca_{0.15}Zr_{0.85}O_{1.85}$ was 0.2%.

### 3.1.2. Textural Characterization of the As–Prepared Sorbents

The $N_2$–adsorption/desorption isotherms and the corresponding pore–size distribution (BJH method) of as–prepared sorbents are shown in Figure 2, and the main textural properties are compiled in Table 1. According to the empirical classification given by the International Union of Pure and Applied Chemistry (IUPAC) [41], CZ belonged to a Type II isotherm (Figure 2a) of macroporous material, characterized by unrestricted multilayer adsorption. At pressure higher than $p/p°$ ~0.8, the absorption and desorption isotherms showed nearly vertical shapes with a negligible hysteresis loop, assigned to the condensation on mesopore and macropore. The pore–size distribution (PSD) calculated using the BJH method (Figure 2b) showed a weak curve with a maximum around ~710 Å assigned to macropores. The specific surface area (SSA) was 11.2 $m^2 \cdot g^{-1}$ and the pore volume was 0.043 $cm^3 \cdot g^{-1}$.

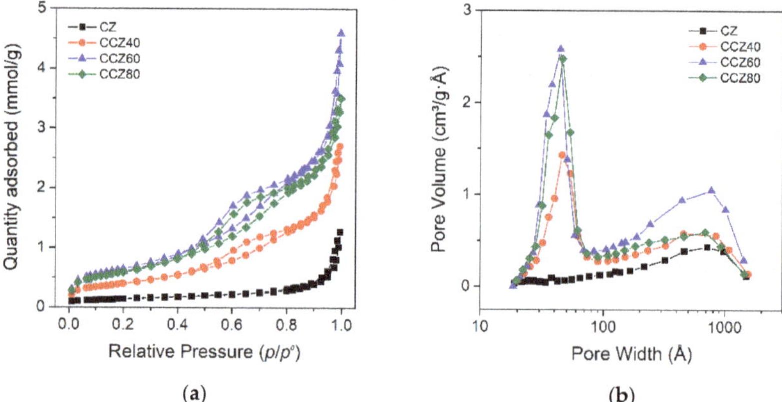

**Figure 2.** Textural characterization of as–prepared samples: (**a**) $N_2$ physisorption isotherms; (**b**) pore–size distribution by the BJH method.

The $N_2$–absorption/desorption isotherms of the CCZ samples were significantly different from those of the CZ perovskite. They were classified as a Type IV isotherm (Figure 2a), given by mesoporous materials, that possess a hysteresis loop. The initial part of the absorption isotherms increased up to the low pressure of $p/p^0 < 0.1$, due to monolayer–multilayer adsorption as in the case of the Type II isotherm. The lower and upper branches of the hysteresis loop were in the range of $p/p^0 = 0.5$–0.8 and the hysteresis was classified, accordingly to IUPAC, as type H1 associated to porous materials with cylindrical–like pores or agglomerates of approximately uniform spheres. As in the case of the CZ sample, at pressure higher than $p/p^0$ ~0.8, the absorption and desorption isotherms showed nearly vertical shapes due to the condensation on macropores and mesopores, suggesting also the presence of a Type II isotherm part. The PSD calculated using the BJH method (Figure 2b) showed two peaks, one at around ~44 Å assigned to the mesopores, and another with a maximum around ~710 Å assigned to the macropores.

The SSA increased from 32.4 m$^2$·g$^{-1}$ to 52.7 m$^2$·g$^{-1}$ on the CCZ40 and CCZ60 samples, respectively, whereas it decreased to 48.0 m$^2$·g$^{-1}$ in the CCZ80 sample with the highest content of calcium, suggesting that the progressive amount of the Ca(OH)$_2$ and CaO species did not continuously increase the SSA of the oxides. Accordingly, the pore volume also showed a similar trend that increased from 0.095 cm$^3$·g$^{-1}$ up to 0.160 cm$^3$·g$^{-1}$ on the CCZ40 and CCZ60 samples, respectively, but decreased to 0.123 cm$^3$·g$^{-1}$ on the CCZ80 one. It is noteworthy that the CZ perovskite shows a unimodal distribution of pores with the maximum located at 800 Å. When the perovskite is used as spacer, the porosity of the sorbent shows a bimodal distribution—one peak is still located at 800 Å, whereas the second peak is at 50 Å. The pore volume increases with the increase of CaO content.

### 3.1.3. SEM Images of the As–Prepared Sorbents

The characteristic morphology of the as synthetized sorbents was observed by SEM images and the main results are shown in Figure 3. The CZ powders showed the typical morphology of perovskite synthetized using the auto–combustion method, using glycine as fuel, with aggregated particles having several dimensions with an average size of 227 ± 118 nm, with different shapes, i.e., spherical, elongated (similar to tetragonal), and planar. The addition of calcium, forming the CCZ samples, substantially modified the morphology. The CaZrO$_3$ particles showed a spherical morphology deposited on much larger particles with three–dimensional polygon shapes, partially attached to each other, assigned to CaO and/or Ca(OH)$_2$ species. The number of CaZrO$_3$ particles followed the sample order of CCZ40 > CCZ60 > CCZ80, decreasing, as expected, according to the higher Ca content compared to Zr. On the basis of the XRD analysis, the bigger particles consisted exclusively of Ca(OH)$_2$ in CCZ80, whereas they were formed by Ca(OH)$_2$ and CaO in the other samples. In particular, for CCZ40 the major phase was still constituted by $Ca(OH)_2$, whereas in CCZ60 there was an opposite trend with the CaO phase double with respect to $Ca(OH)_2$. As shown in Figure 3, CCZ40 and CCZ60 showed a similar morphology with partially aggregated particles, whereas the CCZ80 sample showed more particle agglomeration.

The average size of the CaO, $Ca(OH)_2$, and $CaZrO_3$ particles measured on the as–prepared CCZ samples is reported in Figure 4. The average size of the $CaZrO_3$ particles was $79 \pm 27 > 48 \pm 10 < 53 \pm 14$ nm for the CCZ40, CCZ60, and CCZ80 samples, respectively. The size of the bigger particles attributed to CaO and $Ca(OH)_2$ species showed a similar trend. In fact, it was $485 \pm 69$ nm $> 255 \pm 53$ nm $< 366 \pm 99$ nm for the CCZ40, CCZ60, and CCZ80 samples, respectively. Therefore, the particles of all the species first decreased and then increased as the calcium content increased. SEM images appear to be in agreement with BET results, where the largest surface area was found for the CCZ60 sample.

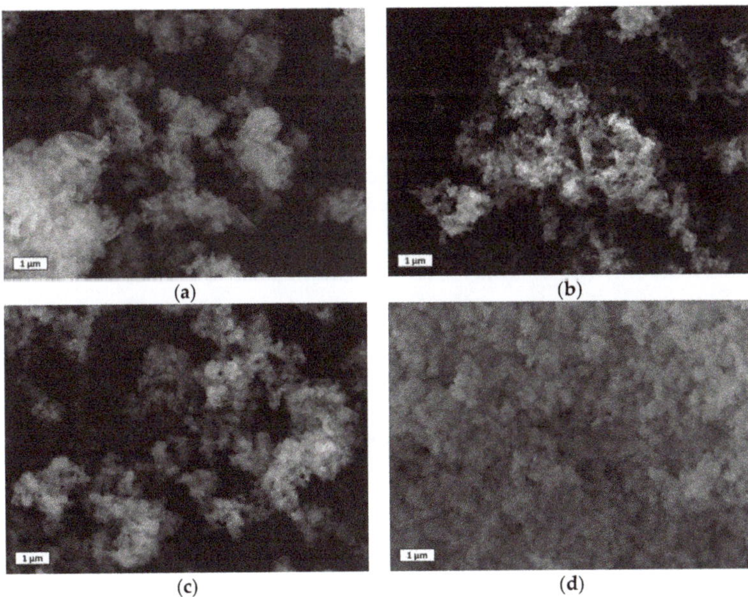

**Figure 3.** Scanning electron microscopy (SEM) images using secondary electron of as–prepared samples: (**a**) CZ, (**b**) CCZ40, (**c**) CCZ60, and (**d**) CCZ80.

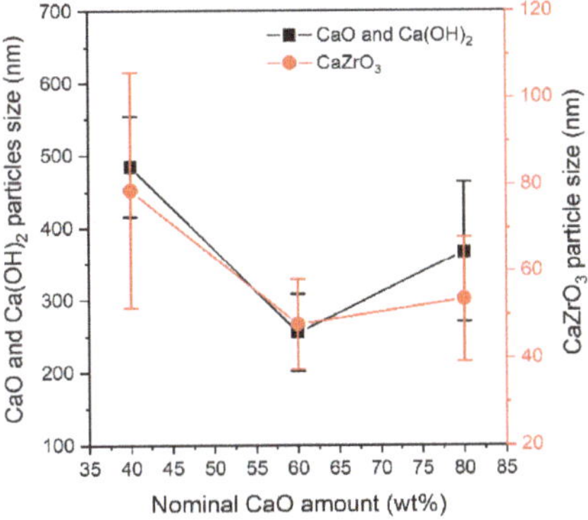

**Figure 4.** Average size of the CaO, $Ca(OH)_2$, and $CaZrO_3$ particles measured by SEM of as–prepared CCZ samples.

### 3.2. $CO_2$ Capture Performance

Figure 5 show the results of the thermogravimetric/differential thermal analysis (TG/DTA) obtained in Ar flow on freshly prepared sorbents before the carbonation and calcination cycles; the results are also summarized in Table 2 In the temperature range 350–500 °C, the TG curves of the CCZ40, CCZ60, and CCZ80 samples showed a weight loss of 5.9%, 13.2%, and 17.6%, respectively, accompanied by a strong endothermic peak reported on the DTA curve, attributed to the decomposition of calcium hydroxide to calcium oxide. At higher temperatures, between 600 and 750 °C, a second endothermic peak was observed, which was associated with a weight loss equal to about 2.3% (CCZ40), 4.1% (CCZ60), and 4.6% (CCZ80) related to the decomposition of calcium carbonate to calcium oxide. For temperatures above 800 °C, there were no variations in weight or heat flows, indicating that the phases of the samples were stable.

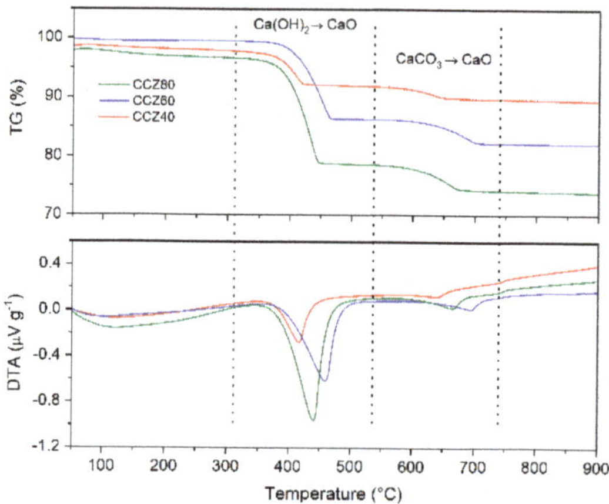


**Figure 5.** Thermogravimetric/differential thermal analysis (TG/DTA) curves of CaO–CaZrO$_3$ at different % of CaO.

**Table 2.** Peak temperature (from differential scanning calorimetry (DSC)) and weight loss (from thermogravimetric analysis (TGA)) of the samples.

| Sample | Ca Species | Peak Temperature (°C) | Weight loss (%) |
|---|---|---|---|
| CZ | CaCO$_3$ | 792.0 | 1.6 |
| CCZ40 | Ca(OH)$_2$ | 417.0 | 5.9 |
|  | CaCO$_3$ | 638.7 | 2.3 |
| CCZ60 | Ca(OH)$_2$ | 458.6 | 13.2 |
|  | CaCO$_3$ | 694.7 | 4.1 |
| CCZ80 | Ca(OH)$_2$ | 440.4 | 17.6 |
|  | CaCO$_3$ | 665.5 | 4.6 |

The cyclic $CO_2$ capacity under mild conditions of the CCZ samples is shown in Figure 6. The $CO_2$ uptake of CCZ40 remained constant for 12 cycles with about 0.31 grams of $CO_2$ uptake per gram of sorbent, a value closer to the theoretical one, which corresponded to almost 100% of CaO conversion. The initial $CO_2$ uptake of the other samples increased with the CaO amount, reaching 0.47 and 0.57 g of $CO_2$ per gram of sorbent for the CCZ60 and CCZ80 samples, respectively. The corresponding initial $CO_2$ conversion was ~100% for CCZ60, whereas it was lower, about ~90%, for the CCZ80 one. Contrary to the stability found for the CCZ40 sample, the other samples showed a decrease of the $CO_2$ uptake per gram of sorbent after 12 cycles. In fact, the final $CO_2$ uptake was 0.34 and 0.42 g of $CO_2$ per gram of sorbent for the CCZ60 and CCZ80 samples, respectively. The calculated loss of CaO conversion was about −27% for both CCZ60 and CCZ80 samples.

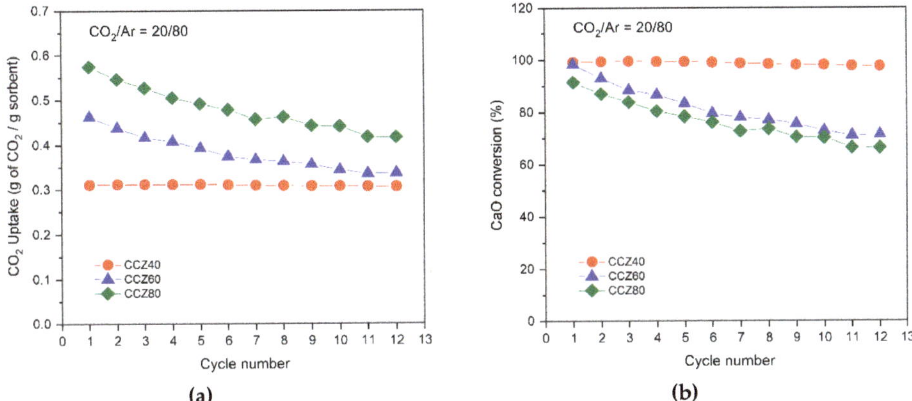

**Figure 6.** Stability and performance tests of CCZ samples under mild conditions showing the $CO_2$ uptake over 12 carbonation/calcination cycles (**a**) and the calculated CaO conversion (**b**).

Given its promising stability, the CCZ40 sample was analyzed under severe conditions. A comparison of the first and last carbonatation steps observed by TG analysis is reported in Figure 7. The carbonation reaction at 650 °C showed a typical two–step profile, i.e., a rapid weight increase, after a few minutes, attributed to the fast gas–solid reaction, followed by a much slower weight increase due to the diffusion resistance of $CO_2$ gas to reach CaO across the as–formed $CaCO_3$ shell. As mentioned before, under mild conditions the weight gain at 650 °C remained very stable. In fact, only a small increase in the diffusion–controlled step with the cycle number was observed. During the temperature ramp in Ar, all cycles showed a rapid decrease in weight that terminated at about 780 °C.

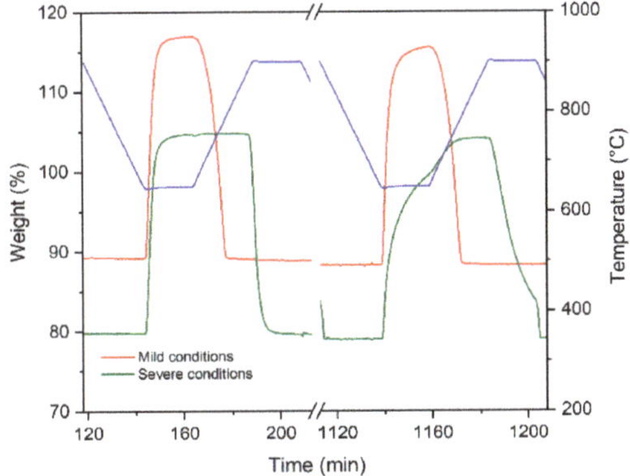

**Figure 7.** TGA weight profiles vs. time of the CCZ40 sorbent under mild conditions (red line) and under severe conditions (green line) for the first and last cycles.

In the isotherms at 650 °C, the first cycle under severe conditions showed a weight increase very similar to that found under mild conditions, with a rapid increase in the mass of $CaCO_3$. Contrary to what was observed under mild conditions, the presence of a high partial pressure of $CO_2$ during the temperature ramp avoided the decomposition of $CaCO_3$, and the mass remained constant up to about

890 °C; when it reached 900 °C, the rapid decarbonation $CaCO_3 \rightarrow CaO + CO_2$ occurred, accompanied by a mass decrease. However, the last cycle showed a significantly different trend. In the isotherm at 650 °C, the formation of $CaCO_3$ decreased markedly due to the increased diffusion resistance. In the subsequent temperature ramp, the formation of $CaCO_3$ increased rapidly again, because it was kinetically favored, returning to a similar weight gain. The $CO_2$ uptake and the corresponding CaO conversion (%) during 12 cycles under severe conditions are reported in Figure 8. In carbonation isotherms at 650 °C, a decrease in performance was observed and the final $CO_2$ uptake and the CaO conversion (%) were equal to 0.26 grams of $CO_2$ per gram of sorbent and 82%, respectively, with a deactivation of 18%. However, it should be noted that after four cycles the activity decayed very slowly, tending to stabilize. Furthermore, it was noted that in the subsequent temperature ramps there was a greater overall stability as shown by the stable $CO_2$ uptake and CaO conversion at the representative temperature of 850 °C, even at a higher $CO_2$ concentration of 80 vol%.

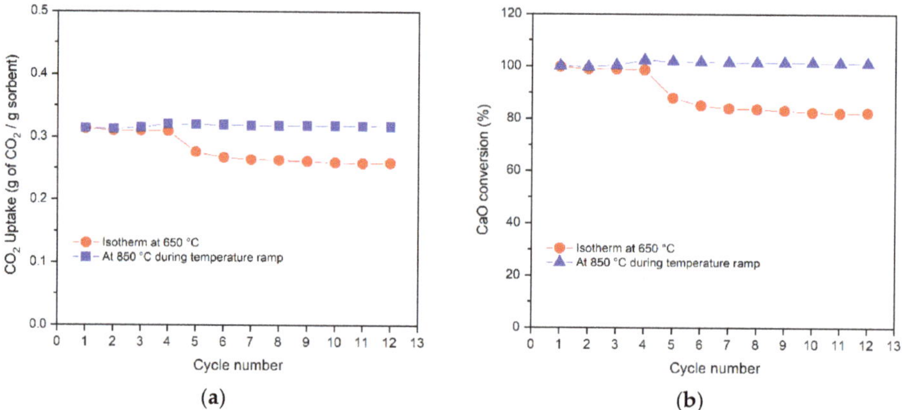



**Figure 8.** Stability and performance tests of CCZ40 sample under severe conditions showing the $CO_2$ uptake over 12 cycles (**a**) and the calculated CaO conversion (**b**) in the isotherm step at 650 °C (red) and at 850 °C during the temperature ramp up to 900 °C (blue).

### 3.3. Shrinking Core Model

The gas–solid reaction for CaO carbonatation was analyzed according to the well–known shrinking core spherical grain model described in Equation (5) in its derivative form and in Equation (6) in its integral form. The variable $X$ is the conversion of CaO to $CaCO_3$ and $r^0$ is the grain model reaction rate. When the reaction is under the kinetic regime, the diffusion effects are negligible, and Equation (5) is represented by a straight line with slope $r^0$:

$$\frac{dX}{dt(1-X)^{\frac{2}{3}}} = 3r^0 \tag{5}$$

$$\left[1 - (1-X)^{\frac{1}{3}}\right] = r^0 \times t \tag{6}$$

The $\left[1 - (1-X)^{\frac{1}{3}}\right]$ against time curves together with the straight line representing the apparent kinetic regime for the first and last carbonatation steps under mild and severe conditions are reported in Figure 9. At the beginning of the experiments, the solid sample can be regarded as an agglomerate of non–porous grains reacting in absence of both extra–particle and intra–particle diffusional effects. Thus, the progress of the reaction, $X$, in the investigated sample can be examined by the shrinking core spherical grain model. As expected in the first cycle, a reaction rate $r^0$, corresponding to 0.12 and 0.13 min$^{-1}$, was obtained under mild and severe conditions, respectively. After 12 cycles under mild conditions, the reaction rate did not decrease and $r^0$ remained as the same value as the first one, i.e., 0.12 min$^{-1}$. The linear phase seems to hold within the first four minutes of carbonation when the sorbent is calcined under mild conditions. Conversely, operating under severe conditions entailed a significant decrease in the reaction rate $r^0$ to 0.06 min$^{-1}$. The linear behavior of carbonation holds in the first three minutes when the sorbent is regenerated under sever conditions. As the carbonation proceeds, the solid reagent is coated by a non–porous carbonate layer. The carbon dioxide is transported via solid state diffusion and the overall process can be controlled by the slower diffusion process, and the reaction intensity decreases (see Figure 9).

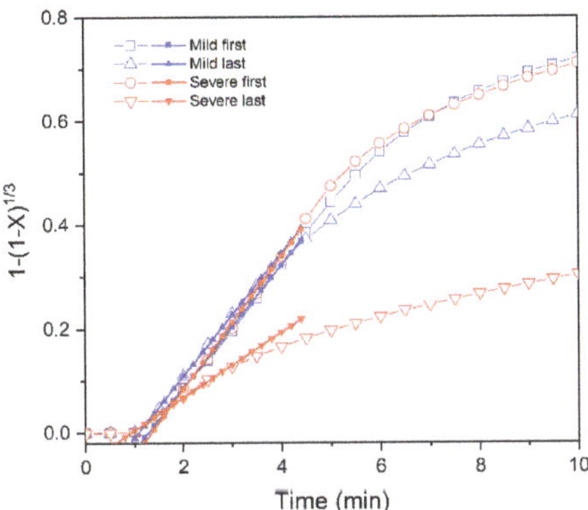


**Figure 9.** CaO carbonatation kinetic curves of the CCZ40 sorbent for the first and last cycles under mild conditions (blue) and under severe conditions (red) and the corresponding slope extraction according to the grain model.

### 3.4. SEM Images of Used Sorbents

The characteristic morphology of the used sorbents was observed by SEM images, and the main results are shown in Figure 10. The CA sample, after the $CO_2$ capture and release cycles, showed characteristic morphologies, also detected in the synthesized sample. However, the main morphology corresponded to aggregated particles of elongated and planar shape. The average length was 371 ± 129 nm, greater than that measured on the as–prepared samples, due to the overall particle sintering. The CCZ samples, after the calcium looping cycles, showed a significantly different morphology than that of the as–prepared samples. The amount of $CaZrO_3$ particles decreased by increasing the CaO content, as expected, maintaining a similar spherical morphology in all the samples similarly to that of the freshly prepared samples. The particles were mainly attached to those of CaO in the CCZ60 and CCZ80 samples, while many particles were also dispersed and separated from those of CaO in the CCZ40 sample, which had the greatest amount of $CaZrO_3$. After the calcium looping cycles, the main morphological difference was found in the particles constituted by the $Ca(OH)_2$ and CaO phases. In fact, they became CaO particles with significantly larger dimensions with respect to the freshly prepared samples and with a main morphology of spherical or even cubic and tetragonal particles.

The average size of the CaO and $CaZrO_3$ particles measured after calcium looping cycles on the CCZ samples is reported in Figure 11. As noted also in the freshly prepared CCZ samples, the size of the $CaZrO_3$ particles had an inverse volcano trend in which the particles had the dimensions 79 ± 16 > 51 ± 9 < 66 ± 14 nm in the CCZ40, CCZ60, and CCZ80 samples, respectively. It is interesting to note that the dimensions measured both in the fresh samples and after the calcium looping cycles were similar, suggesting that the $CaZrO_3$ particles were stable towards sintering. On the contrary, the CaO particles were significantly larger than those of CaO and $Ca(OH)_2$ present in the fresh samples.

The particle size of CaO increased proportionally to the Ca content in the samples, being 525 ± 101 < 588 ± 166 < 604 ± 137 nm in the CCZ40, CCZ60, and CCZ80 samples, respectively.

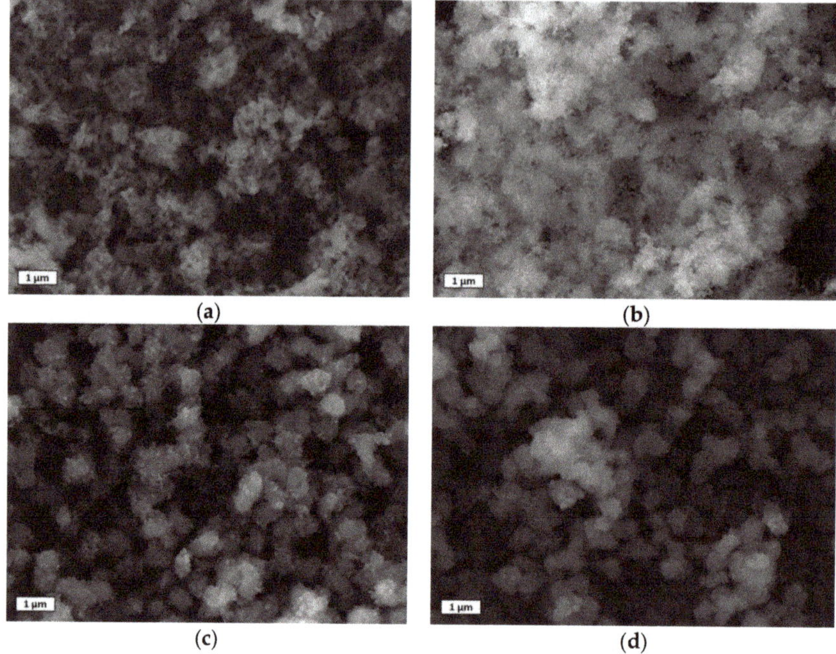

Figure 10. SEM images using secondary electron of used samples: (a) CZ, (b) CCZ40, (c) CCZ60, and (d) CCZ80.

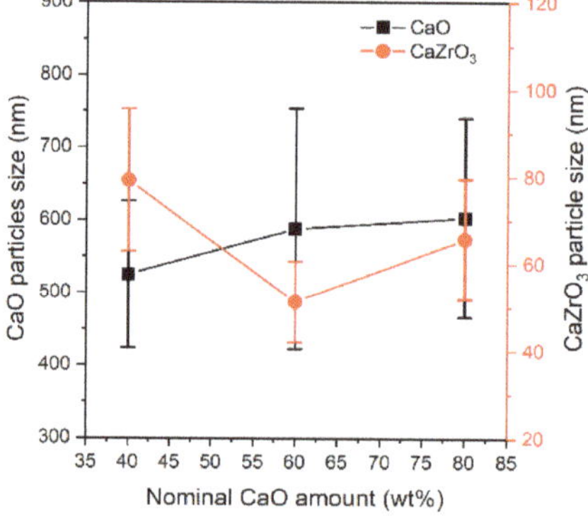

**Figure 11.** Average size of the CaO and $CaZrO_3$ particles measured by SEM after calcium looping cycles of the CCZ samples.

## 4. Conclusions

$CaZrO_3$ was introduced into CaO–based sorbents to increase stability during repeated $CO_2$ capture/release cycles. The CaO–$CaZrO_3$ sorbents with different CaO contents (40, 60, and 80 wt%) were synthesized using the self–combustion method. The calcium looping process was performed under mild and severe conditions. Under mild conditions (decarbonation in pure Ar), the $CO_2$

capture capacity increased with the CaO content. After 12 cycles, the most stable sorbent was the one with 40 wt% of CaO (CCZ40), which maintained a CaO conversion to $CaCO_3$ nearly 100% and a $CO_2$ uptake of 0.31 g $CO_2$/g$_{sorbent}$. The CCZ40 sample showed high performance even operating under severe conditions (decarbonation under 85% $CO_2$) with a decrease of the $CO_2$ uptake from 0.31 g $CO_2$/g$_{sorbent}$ to 0.26 g $CO_2$/g$_{sorbent}$. This experimental result confirms the negligible diffusional resistance of $CO_2$ throughout the particle and the good gas–solid contacting even at the inner core of the sorbent particle. SEM images of the used sorbent showed larger micrometric CaO aggregates surrounded by nanometric $CaZrO_3$ particles. The best stability was attributed to the correct balance between CaO, the active component, and the $CaZrO_3$ nanoparticles. The experimental data gathered from the thermogravimetric analyzer corroborated the adoption of the shrinking core spherical model for the interpretation of CaO conversion to $CaCO_3$. A maximum reaction rate of 0.12–0.13 $min^{-1}$ was evaluated during carbonation in the multi–cycling $CO_2$ capture under mild conditions. The reaction rate evaluated at the 12th cycle decreased to 0.06 $min^{-1}$ when the sorbent was regenerated under severe conditions.

**Author Contributions:** Conceptualization, I.L. and S.S.; materials synthesis, I.L.; XRD analysis, R.C.; BET analysis, I.L. and G.V.; TG–DTA analysis, M.R.M.; SEM analysis, L.D.S.; writing—original draft preparation, I.L. and S.S.; writing—review and editing the final version, I.L., S.S., R.C., and M.L.G. All authors have read and agreed to the published version of the manuscript.

**Funding:** This research was funded by the Italian Ministry of Economic Development, under the project SFERO (Systems for Flexible Energy via Reuse of carbOn), grant number B09Y.

**Conflicts of Interest:** The authors declare no conflict of interest.

## References

1.  Abanades, J.C.; Anthony, E.J.; Lu, D.Y.; Salvador, C.; Alvarez, D. Capture of $CO_2$ from combustion gases in a fluidized bed of CaO. *AIChE J.* **2004**, *50*, 1614–1622. [CrossRef]

2.  Bhatia, S.K.; Perlmutter, D.D. Effect of the product layer on the kinetics of the $CO_2$–lime reaction. *AIChE J.* **1983**, *29*, 79–86. [CrossRef]

3.  Stendardo, S.; Di Felice, L.; Gallucci, K.; Foscolo, P.U. $CO_2$ capture with calcined dolomite: The effect of sorbent particle size. *Biomass Convers. Biorefinery* **2011**, *1*, 149–161. [CrossRef]

4.  Abanades, J.C.; Anthony, E.J.; Wang, J.; Oakey, J.E. Fluidized Bed Combustion Systems Integrating $CO_2$ Capture with CaO. *Environ. Sci. Technol.* **2005**, *39*, 2861–2866. [CrossRef] [PubMed]

5.  Zhao, M.; He, X.; Ji, G.; Song, Y.; Zhao, X. Zirconia incorporated calcium looping absorbents with superior sintering resistance for carbon dioxide capture from in situ or ex situ processes. *Sustain. Energy Fuels* **2018**, *2*, 2733–2741. [CrossRef]

6.  Herce, C.; Cortés, C.; Stendardo, S. Computationally efficient CFD model for scale–up of bubbling fluidized bed reactors applied to sorption–enhanced steam methane reforming. *Fuel Process. Technol.* **2017**, *167*, 747–761. [CrossRef]

7.  Wang, W.; Li, Y.; Xie, X.; Sun, R. Effect of the presence of HCl on cyclic CO2 capture of calcium–based sorbent in calcium looping process. *Appl. Energy* **2014**, *125*, 246–253. [CrossRef]

8.  Stendardo, S.; Foscolo, P.U. Carbon dioxide capture with dolomite: A model for gas–solid reaction within the grains of a particulate sorbent. *Chem. Eng. Sci.* **2009**, *64*, 2343–2352. [CrossRef]

9.  Stendardo, S.; Andersen, L.K.; Herce, C. Self–activation and effect of regeneration conditions in CO2–carbonate looping with CaO–Ca12Al14O33 sorbent. *Chem. Eng. J.* **2013**, *220*, 383–394. [CrossRef]

10.  Berger, E.E. Effect of Steam on the Decomposition of Limestone1,1. *Ind. Eng. Chem.* **1927**, *19*, 594–596. [CrossRef]

11.  Wang, Y.; Lin, S.; Suzuki, Y. Limestone Calcination with $CO_2$ Capture (II): Decomposition in $CO_2$/Steam and $CO_2$/N$_2$ Atmospheres. *Energy Fuels* **2008**, *22*, 2326–2331. [CrossRef]

12.  Abanades, J.C.; Rubin, E.S.; Anthony, E.J. Sorbent Cost and Performance in CO2 Capture Systems. *Ind. Eng. Chem. Res.* **2004**, *43*, 3462–3466. [CrossRef]

13.  Gallucci, K.; Stendardo, S.; Foscolo, P.U. $CO_2$ capture by means of dolomite in hydrogen production from syn gas. *Int. J. Hydrogen Energy* **2008**, *33*, 3049–3055. [CrossRef]

14. Fennell, P.S.; Pacciani, R.; Dennis, J.S.; Davidson, J.F.; Hayhurst, A.N. The Effects of Repeated Cycles of Calcination and Carbonation on a Variety of Different Limestones, as Measured in a Hot Fluidized Bed of Sand. *Energy Fuels* **2007**, *21*, 2072–2081. [CrossRef]

15. Sun, P.; Grace, J.R.; Lim, C.J.; Anthony, E.J. The effect of CaO sintering on cyclic CO2 capture in energy systems. *AIChE J.* **2007**, *53*, 2432–2442. [CrossRef]

16. Fang, F.; Li, Z.-S.; Cai, N.-S. Experiment and Modeling of $CO_2$ Capture from Flue Gases at High Temperature in a Fluidized Bed Reactor with Ca–Based Sorbents. *Energy Fuels* **2009**, *23*, 207–216. [CrossRef]

17. Manovic, V.; Anthony, E.J. Thermal Activation of CaO–Based Sorbent and Self–Reactivation during CO2 Capture Looping Cycles. *Environ. Sci. Technol.* **2008**, *42*, 4170–4174. [CrossRef]

18. Chen, Z.; Song, H.S.; Portillo, M.; Lim, C.J.; Grace, J.R.; Anthony, E.J. Long–Term Calcination/Carbonation Cycling and Thermal Pretreatment for CO2 Capture by Limestone and Dolomite. *Energy Fuels* **2009**, *23*, 1437–1444. [CrossRef]

19. Lysikov, A.I.; Salanov, A.N.; Okunev, A.G. Change of $CO_2$ Carrying Capacity of CaO in Isothermal Recarbonation–Decomposition Cycles. *Ind. Eng. Chem. Res.* **2007**, *46*, 4633–4638. [CrossRef]

20. Hughes, R.W.; Lu, D.; Anthony, E.J.; Wu, Y. Improved Long–Term Conversion of Limestone–Derived Sorbents for In Situ Capture of CO2 in a Fluidized Bed Combustor. *Ind. Eng. Chem. Res.* **2004**, *43*, 5529–5539. [CrossRef]

21. Rodríguez, N.; Alonso, M.; Abanades, J.C. Experimental investigation of a circulating fluidized–bed reactor to capture $CO_2$ with CaO. *AIChE J.* **2011**, *57*, 1356–1366. [CrossRef]

22. Arias, B.; Grasa, G.S.; Abanades, J.C. Effect of sorbent hydration on the average activity of CaO in a Ca–looping system. *Chem. Eng. J.* **2010**, *163*, 324–330. [CrossRef]

23. Borgwardt, R.H. Calcium oxide sintering in atmospheres containing water and carbon dioxide. *Ind. Eng. Chem. Res.* **1989**, *28*, 493–500. [CrossRef]

24. Beruto, D.; Barco, L.; Searcy, A.W. $CO_2$–Catalyzed Surface Area and Porosity Changes in High–Surface–Area CaO Aggregates. *J. Am. Ceram. Soc.* **1984**, *67*, 512–516. [CrossRef]

25. Ewing, J.; Beruto, D.; Searcy, A.W. The Nature of CaO Produced by Calcite Powder Decomposition in Vacuum and in CO2. *J. Am. Ceram. Soc.* **1979**, *62*, 580–584. [CrossRef]

26. Zeman, F. Effect of steam hydration on performance of lime sorbent for CO2 capture. *Int. J. Greenh. Gas Control* **2008**, *2*, 203–209. [CrossRef]

27. Wang, K.; Guo, X.; Zhao, P.; Zhang, L.; Zheng, C. CO2 capture of limestone modified by hydration–dehydration technology for carbonation/calcination looping. *Chem. Eng. J.* **2011**, *173*, 158–163. [CrossRef]

28. Meis, N.N.A.H.; Bitter, J.H.; de Jong, K.P. Support and Size Effects of Activated Hydrotalcites for Precombustion CO2 Capture. *Ind. Eng. Chem. Res.* **2010**, *49*, 1229–1235. [CrossRef]

29. Reijers, H.T.J.; Valster–Schiermeier, S.E.A.; Cobden, P.D.; van den Brink, R.W. Hydrotalcite as $CO_2$ Sorbent for Sorption–Enhanced Steam Reforming of Methane. *Ind. Eng. Chem. Res.* **2006**, *45*, 2522–2530. [CrossRef]

30. Herce, C.; Stendardo, S.; Cortés, C. Increasing $CO_2$ carrying capacity of dolomite by means of thermal stabilization by triggered calcination. *Chem. Eng. J.* **2015**, *262*, 18–28. [CrossRef]

31. Gupta, H.; Fan, L.-S. Carbonation–Calcination Cycle Using High Reactivity Calcium Oxide for Carbon Dioxide Separation from Flue Gas. *Ind. Eng. Chem. Res.* **2002**, *41*, 4035–4042. [CrossRef]

32. Lu, H.; Smirniotis, P.G. Calcium Oxide Doped Sorbents for $CO_2$ Uptake in the Presence of SO2 at High Temperatures. *Ind. Eng. Chem. Res.* **2009**, *48*, 5454–5459. [CrossRef]

33. Lu, H.; Khan, A.; Smirniotis, P.G. Relationship between Structural Properties and CO2 Capture Performance of CaO–Based Sorbents Obtained from Different Organometallic Precursors. *Ind. Eng. Chem. Res.* **2008**, *47*, 6216–6220. [CrossRef]

34. Roesch, A.; Reddy, E.P.; Smirniotis, P.G. Parametric Study of Cs/CaO Sorbents with Respect to Simulated Flue Gas at High Temperatures. *Ind. Eng. Chem. Res.* **2005**, *44*, 6485–6490. [CrossRef]

35. Chen, H.; Zhao, C.; Duan, L.; Liang, C.; Liu, D.; Chen, X. Enhancement of reactivity in surfactant–modified sorbent for $CO_2$ capture in pressurized carbonation. *Fuel Process. Technol.* **2011**, *92*, 493–499. [CrossRef]

36. Salvador, C.; Lu, D.; Anthony, E.J.; Abanades, J.C. Enhancement of CaO for $CO_2$ capture in an FBC environment. *Chem. Eng. J.* **2003**, *96*, 187–195. [CrossRef]

37. Li, Z.-S.; Cai, N.-S.; Huang, Y.-Y. Effect of Preparation Temperature on Cyclic $CO_2$ Capture and Multiple Carbonation–Calcination Cycles for a New Ca–Based $CO_2$ Sorbent. *Ind. Eng. Chem. Res.* **2006**, *45*, 1911–1917. [CrossRef]

38. Pacciani, R.; Müller, C.R.; Davidson, J.F.; Dennis, J.S.; Hayhurst, A.N. Synthetic Ca–based solid sorbents suitable for capturing $CO_2$ in a fluidized bed. *Can. J. Chem. Eng.* **2008**, *86*, 356–366. [CrossRef]

39. Wu, S.F.; Zhu, Y.Q. Behavior of CaTiO3/Nano–CaO as a $CO_2$ Reactive Adsorbent. *Ind. Eng. Chem. Res.* **2010**, *49*, 2701–2706. [CrossRef]

40. Hubbard, C.R.; Snyder, R.L. RIR—Measurement and Use in Quantitative XRD. *Powder Diffr.* **2013**, *3*, 74–77. [CrossRef]

41. Sing, K.S.W. Reporting physisorption data for gas/solid systems with special reference to the determination of surface area and porosity (Recommendations 1984). *Pure Appl. Chem.* **1985**, *57*, 603–619. [CrossRef]

MDPI

St. Alban-Anlage 66

4052 Basel

Switzerland

Tel. +41 61 683 77 34

Fax +41 61 302 89 18

www.mdpi.com

*Metals* Editorial Office

E-mail: metals@mdpi.com

www.mdpi.com/journal/metals